TWELFTH EDITION

BIOLOGY

FOCUS REVIEW GUIDE
for AP® Biology

Mc
Graw
Hill
Education

About the AP Consultant

Darrel James received his BS in Biology from Pacific University in Oregon and his MS in Marine Science from Oregon State. He taught Biology for thirty years, and is presently teaching at Beyer High School in Modesto, CA. Darrel teaches Pre-AP as well as AP Biology and is the department chairperson. Darrel has served as a reader, table leader, and assistant chief reader for the AP Biology exam since 1990. He has led numerous one-day AP Biology curriculum and grading workshops for the College Board, and he has presented week-long institutes for the past twenty-four summers. Darrel is a member of the California Commission on Science and Technology. He is the vice-chair of the California Teachers Advisory Council that is an arm of the CCST which advises the state on STEM and digitally enhanced education. Darrel has led marine biology trips to Hawaii for his AP Biology students. He is also an avid cyclist. He loves to teach and show people how much fun biology can be.

MHEonline.com

Send all inquiries to:
McGraw-Hill Education
8787 Orion Place
Columbus, OH 43240

ISBN: 978-0-07-672152-8
MHID: 0-07-672152-3

Printed in the United States of America.

2 3 4 5 6 7 8 9 QTN 18 17 16 15

Table of Contents

Using Your AP Focus Review Guide

This review guide was developed with the AP student in mind. The activities within each chapter will help you to focus on and review the key content in the chapter as it relates to the AP Biology Curriculum.

The **Following the Big Ideas** boxes review the Big Ideas identified in your textbook.

These correlations pinpoint which parts of the AP Curriculum are reviewed in each section.

Review It activities will help you to recall content that you should have mastered in earlier study.

You can use this column to take notes and keep track of content you may need to review in preparation for the exam.

Use It activities allow you to apply what you've learned in the chapter.

In the **Summarize It** section you will synthesize what you have learned by focusing in on the key content in the chapter.

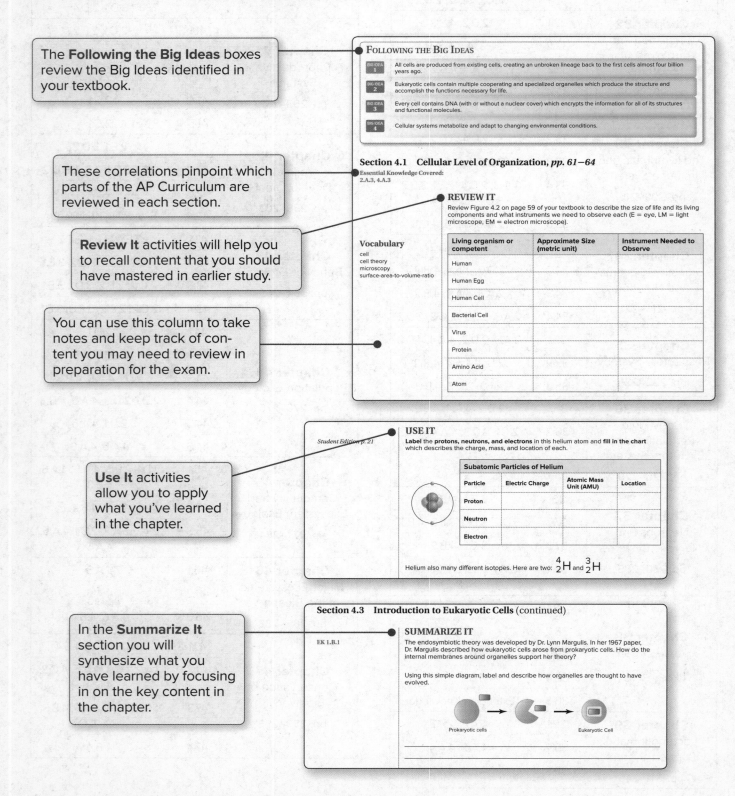

FOLLOWING THE BIG IDEAS

BIG IDEA 1 — All cells are produced from existing cells, creating an unbroken lineage back to the first cells almost four billion years ago.

BIG IDEA 2 — Eukaryotic cells contain multiple cooperating and specialized organelles which produce the structure and accomplish the functions necessary for life.

BIG IDEA 3 — Every cell contains DNA (with or without a nuclear cover) which encrypts the information for all of its structures and functional molecules.

BIG IDEA 4 — Cellular systems metabolize and adapt to changing environmental conditions.

Section 4.1   Cellular Level of Organization, pp. 61–64

Essential Knowledge Covered:
2.A.3, 4.A.3

REVIEW IT

Review Figure 4.2 on page 59 of your textbook to describe the size of life and its living components and what instruments we need to observe each (E = eye, LM = light microscope, EM = electron microscope).

Vocabulary

cell
cell theory
microscopy
surface-area-to-volume-ratio

Living organism or competent	Approximate Size (metric unit)	Instrument Needed to Observe
Human		
Human Egg		
Human Cell		
Bacterial Cell		
Virus		
Protein		
Amino Acid		
Atom		

USE IT

Student Edition p. 21

Label the **protons, neutrons, and electrons** in this helium atom and **fill in the chart** which describes the charge, mass, and location of each.

Subatomic Particles of Helium

Particle	Electric Charge	Atomic Mass Unit (AMU)	Location
Proton			
Neutron			
Electron			

Helium also many different isotopes. Here are two: $^{4}_{2}H$ and $^{3}_{2}H$

Section 4.3   Introduction to Eukaryotic Cells (continued)

SUMMARIZE IT

EK 1.B.1

The endosymbiotic theory was developed by Dr. Lynn Margulis. In her 1967 paper, Dr. Margulis described how eukaryotic cells arose from prokaryotic cells. How do the internal membranes around organelles support her theory?

Using this simple diagram, label and describe how organelles are thought to have evolved.

Prokaryotic cells → → Eukaryotic Cell

Section content that is not a focus of the AP Biology Curriculum is identified as either **Prerequisite Knowledge** or **Extending Knowledge.**

Different visual organizers help you to analyze and summarize information and remember content.

The **AP Focus Review Guide** gives you an opportunity to answer the **Reviewing the Essential Questions** from your textbook, providing an excellent final review of the chapter content.

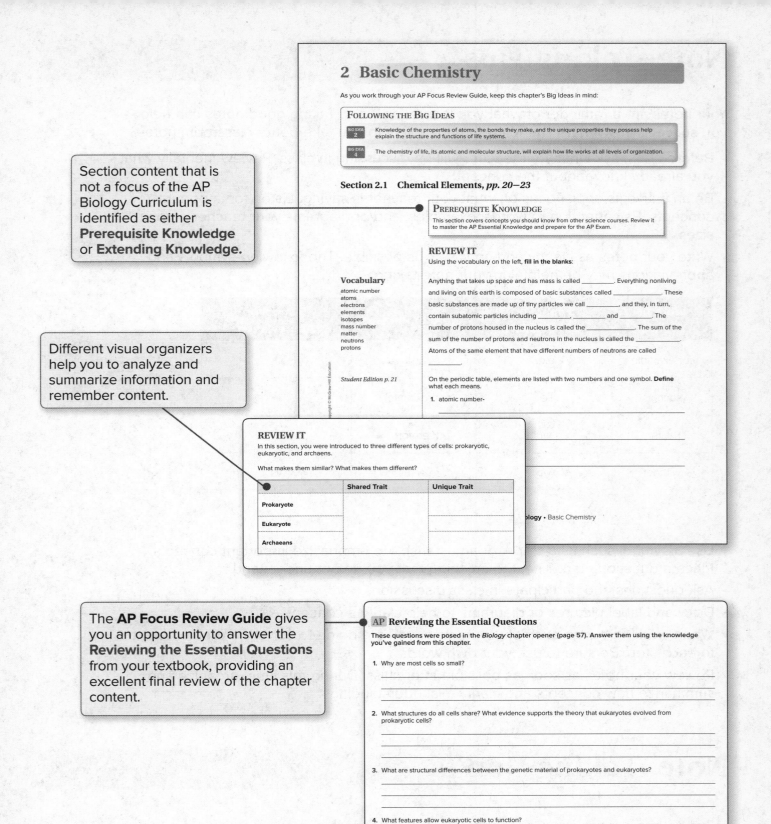

2 Basic Chemistry

As you work through your AP Focus Review Guide, keep this chapter's Big Ideas in mind:

FOLLOWING THE BIG IDEAS

| BIG IDEA 2 | Knowledge of the properties of atoms, the bonds they make, and the unique properties they possess help explain the structure and functions of life systems. |
| BIG IDEA 4 | The chemistry of life, its atomic and molecular structure, will explain how life works at all levels of organization. |

Section 2.1   Chemical Elements, *pp. 20—23*

PREREQUISITE KNOWLEDGE

This section covers concepts you should know from other science courses. Review it to master the AP Essential Knowledge and prepare for the AP Exam.

REVIEW IT

Using the vocabulary on the left, **fill in the blanks**:

Vocabulary
atomic number
atoms
electrons
elements
isotopes
mass number
matter
neutrons
protons

Student Edition p. 21

Anything that takes up space and has mass is called _____. Everything nonliving and living on this earth is composed of basic substances called _____. These basic substances are made up of tiny particles we call _____, and they, in turn, contain subatomic particles including _____, _____ and _____. The number of protons housed in the nucleus is called the _____. The sum of the sum of the number of protons and neutrons in the nucleus is called the _____. Atoms of the same element that have different numbers of neutrons are called _____.

On the periodic table, elements are listed with two numbers and one symbol. **Define** what each means.

1. atomic number-
    _____

REVIEW IT

In this section, you were introduced to three different types of cells: prokaryotic, eukaryotic, and archaens.

What makes them similar? What makes them different?

	Shared Trait	Unique Trait
Prokaryote		
Eukaryote		
Archaeans		

logy • Basic Chemistry

AP Reviewing the Essential Questions

These questions were posed in the *Biology* chapter opener (page 57). Answer them using the knowledge you've gained from this chapter.

1. Why are most cells so small?
    _____
    _____
    _____

2. What structures do all cells share? What evidence supports the theory that eukaryotes evolved from prokaryotic cells?
    _____
    _____
    _____

3. What are structural differences between the genetic material of prokaryotes and eukaryotes?
    _____
    _____
    _____

4. What features allow eukaryotic cells to function?
    _____
    _____
    _____

Note–Taking Tips

Your notes are a reminder of what you learned in class. Taking good notes can help you succeed in science. The following tips will help you take better classroom notes.

- Before class, ask what your teacher will be discussing in class. Review mentally what you already know about the concept.
- Be an active listener. Focus on what your teacher is saying. Listen for important concepts. Pay attention to words, examples, and/or diagrams your teacher emphasizes.
- Write your notes as clearly and concisely as possible. The following symbols and abbreviations may be helpful in your note-taking.

Word or Phrase	Symbol or Abbreviation	Word or Phrase	Symbol or Abbreviation
for example	e.g.	and	+
such as	i.e.	approximately	≈
with	w/	therefore	∴
without	w/o	versus	vs

- Use a symbol such as a star ★ or an asterisk * to emphasize important concepts. Place a question mark ? next to anything that you do not understand.
- Ask questions and participate in class discussion.
- Draw and label pictures or diagrams to help clarify a concept.
- When working out an example, write what you are doing to solve the problem next to each step. Be sure to use your own words.
- Review your notes as soon as possible after class. During this time, organize and summarize new concepts and clarify misunderstandings.

Note–Taking Don'ts

- **Don't** write every word. Concentrate on the main ideas and concepts.
- **Don't** use someone else's notes. They may not make sense.
- **Don't** doodle. It distracts you from listening actively.
- **Don't** lose focus or you will become lost in your note-taking.

1 A View of Life

As you work through your AP Focus Review Guide, keep this chapter's Big Ideas in mind:

FOLLOWING THE BIG IDEAS

 BIG IDEA 1 Understanding the scientific process, the theory of evolution, and the interactions of biological systems is important in the study of biology.

 BIG IDEA 2 Living organisms obey the same laws of chemistry and physics that govern everything within the universe but are distinguished by characteristics unique to life.

 BIG IDEA 3 Living organisms detect and respond to changes in their environment and pass information to other organisms in their community.

BIG IDEA 4 For communities of individual cells to organisms, all life is based on atoms and molecules.

Section 1.1 Introduction to AP Biology, *p. 2*

In your own words, what is the field of biology all about?

_____

_____

In the chart below, describe the attributes of living things.

Notes

Life- basic characteristics:
- *cells*
- *genes-DNA*
- *common ancestor*

Place a star next to the category that all living things have in common.

Size	
Locations	
Life span	
Composition	

Section 1.2 Big Idea 1: Evolution, *pp. 2–5*

Notes

Evolution- Core Concept #1
- *Natural selection*
-
-

How do the fields of taxonomy and systematics relate to evolution?

_____

_____

Organizing Life- Trees
-
-

List the basic classification of organisms going from most inclusive to least inclusive:

→　　　　→　　　　→

　　　　　　　　　　↓

←　　　　←　　　　←

Section 1.3   Big Idea 2: Energy and Molecular Building Blocks, *pp. 6–7*

Notes

Energy-
- *metabolism*
- *photosynthesis*
-

Homeostasis
-
-
-

Draw a simplistic diagram to describe how chemicals and energy flow from the sun to plants to organisms that eat the plants.

Section 1.4   Big Idea 3: Information Storage, Transmission, and Response, *pp. 7–8*

Notes

Organisms
- *Respond*
- *reproduce*

List five things all living things have in common.

Section 1.5   Big Idea 4: Interdependent Relationships, *pp. 8–10*

Notes

Emergent Properties
-
-

Cooperation & Competition
-

Diversity
-

In the chart below, briefly **describe** how the two terms interact.

Organisms		Ecosystems
Cells		Organs
Autotrophs		Heterotrophs

Notes

Science Practices

1. **Models**
   -*illustrate ideas*
   -
2.
   -
3.
   -
4.
   -
5.
   -
6.
   -
7.
   -
   -

Describe the seven AP science practices (SP) that will help you understand biology and how scientists approach their work.

SP1	
SP2	
SP3	
SP4	
SP5	
SP6	
SP7	

Notes

Scientific Method

- **Observation**
- **Hypothesis**
- **Design**
- **Statistics**
-

Find Figure 1.13 in your textbook. The graph at the bottom shows how effective a particular treatment was against a disease in three different groups. Using this graph, answer some questions concerning the scientific method.

Why did the scientists include a control group in their experimental design?

_____

_____

Explain what the "T" shaped lines coming off of the bars on the graph indicate.

_____

_____

If the scientist who performed a statistical test on this data found a p value less than 0.05 between Group 1 and Group 2, is there a significant difference between the two groups?

_____

These scientists sent their study and conclusions off to a scientific journal. What happens to their article before it is published?

_____

_____

These questions were posed in the *Biology* chapter opener (page 1). Answer them using the knowledge you've gained from this chapter.

1. Why is evolution a central theme of the biological sciences?

   _____

   _____

   _____

2. What characteristics distinguish living organisms from nonliving things?

   _____

   _____

   _____

3. How do living organisms detect, respond, and pass on information to other organisms?

   _____

   _____

   _____

4. In what ways do living organisms interact with each other and with their environments?

   _____

   _____

   _____

2 Basic Chemistry

As you work through your AP Focus Review Guide, keep this chapter's Big Ideas in mind:

FOLLOWING THE BIG IDEAS

 BIG IDEA 2 Knowledge of the properties of atoms, the bonds they make, and the unique properties they possess help explain the structure and functions of life systems.

 BIG IDEA 4 The chemistry of life, its atomic and molecular structure, will explain how life works at all levels of organization.

Section 2.1 Chemical Elements, *pp. 20–23*

PREREQUISITE KNOWLEDGE

This section covers concepts you should know from other science courses. Review it to master the AP Essential Knowledge and prepare for the AP Exam.

REVIEW IT

Using the vocabulary on the left, **fill in the blanks** below:

Vocabulary

atomic number
atoms
electrons
elements
isotopes
mass number
matter
neutrons
protons

Anything that takes up space and has mass is called _____. Everything nonliving and living on this earth is composed of basic substances called _____. These basic substances are made up of tiny particles we call _____, and they, in turn, contain subatomic particles including _____, _____ and _____. The number of protons housed in the nucleus is called the _____. The sum of the sum of the number of protons and neutrons in the nucleus is called the _____. Atoms of the same element that have different numbers of neutrons are called _____.

Student Edition p. 21

On the periodic table, elements are listed with two numbers and one symbol. **Define** what each means.

 1. atomic number

_____

 2. atomic symbol

_____

 3. atomic mass

_____

Section 2.1 Chemical Elements (continued)

USE IT

Student Edition p. 21

Label the **protons, neutrons, and electrons** in this helium atom and **fill in the chart** which describes the charge, mass, and location of each.

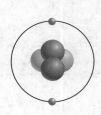

Subatomic Particles of Helium			
Particle	**Electric Charge**	**Atomic Mass Unit (AMU)**	**Location**
Proton			
Neutron			
Electron			

Helium also many different isotopes. Here are two: ^4_2H and ^3_2H

Student Edition p. 21

How many protons do these isotopes have? How many neutrons?

_____

_____

Student Edition p. 22

Some isotopes are highly radioactive. What is radiation?

_____

_____

Student Edition p. 22

Describe one way radiation is harmful and one way it is beneficial to humans.

_____

_____

_____

SUMMARIZE IT

Find the element **boron (B)** on the periodic table on page 22 of your textbook and **answer** the following questions.

a. What is boron's atomic numer?

_____

b. In what group and period is boron located?

_____

c. What is boron's atomic mass?

_____

Section 2.1   Chemical Elements (continued)

A common model used to illustrate electrons in electron shells around the nucleus is called the Bohr model. This model is described on page 23 of your textbook. Following the rules of the Bohr model, **draw** a model of boron. **Label** the nucleus, electrons, and valance shell.

Section 2.2   Molecules and Compounds, *pp. 24–25*

Essential Knowledge Covered:
4.A.1

REVIEW IT

Using the vocabulary on the left, **fill in the blanks** below.

Vocabulary

compound
covalent bonds
electronegativity
electrons
formulas
ionic bonds
ion
polar covalent
polarity
molecules
nonpolar covalent

When elements form covalent bonds, a _____ is formed. The combination of two or more different elements is referred to as a _____. Molecules and compounds are described by _____, which indicate numbers and types of atoms present. Located around the nucleus, _____ play a critical role in the bond formation between elements. An element that has lost or gained an electron from its valence shell is called an _____. _____ are held together by an attraction between negatively and positively charged ions; whereas _____ share electrons to satisfy their valence shell's octet. If you were to measure an atom's attraction for an electron, you would be measuring its _____. Covalent bonds that share electrons equally are referred to as _____, and covalent bonds which share unequally are called _____. Determined by both bond polarity and by molecular shape, the _____ of a molecule affects molecule-molecule interactions.

USE IT

EK 4.A.1
Student Edition p. 24

Students studying for their AP Exams might find themselves imbibing $C_8H_{10}N_4O_2$, caffeine. **Identify** the number of atoms and molecules in this compound.

_____

EK 4.A.1
Student Edition p. 25

Caffeine, $C_8H_{10}N_4O_2$, is made up of many different covalent bonds between its atoms. Given the information that the nitrogen and oxygen atoms present have higher electronegativity compared to the carbon and hydrogen atoms, **determine** whether or not caffeine is polar and describe why.

_____

_____

_____

Section 2.2 Molecules and Compounds (continued)

EK 4.A.1
Student Edition pp. 24—25

Draw an arrow to the bonds which hold these compounds together. **Classify** the bonds as either ionic (I) or covalent (C) in the table below.

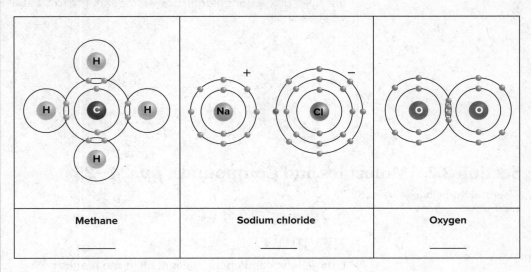

Methane	Sodium chloride	Oxygen
_____	_____	_____

SUMMARIZE IT

EK 4.A.1

Describe the difference between ionic, polar covalent, and nonpolar covalent bonds. Use the following keywords in your description: **ions, electrons,** and **electronegativity.**

_____

_____

_____

_____

EK 4.A.1

Circle the molecule which contains the strongest bond and **explain** your choice.

$$O=O$$
oxygen

$$H-C\equiv C-H$$
acetylene

$$H-H$$
hydrogen

_____

_____

EK 4.A.1

Refer to the diagram of methane on the previous page. **List** two reasons why methane is nonpolar.

1. _____

2. _____

Section 2.3   Chemistry of Water, *pp. 26−29*

Essential Knowledge Covered:
2.A.3, 2.A.3a, 2.A.3a1

REVIEW IT

Water is an amazing molecule. It is not a stretch to say the structure and behavior of water allows life on Earth to exist.

Using the vocabulary, **list and describe** the properties of water that allow each organism in the chart below its certain ability.

Vocabulary

adhesion
cohesion
density
heat of vaporization
hydrogen bonding
surface tension

Organism	Ability	Water Property	How it works
Water strider	Walks across the surface of a pond		
Oak tree	Photosynthesizes 25 m above ground		
Human	Cools off on a hot day by perspiring		
Lake trout	Over-winters in lake that is covered in 10 cm of solid ice		

Section 2.3 Chemistry of Water (continued)

USE IT

EK 2.A.3a
Student Edition p. 27

The graph below shows the energy required to change the temperature of a gram of water. Using the concept illustrated on graph below, **explain** how a camel can live in a desert where temperatures changing from 5 to 40 °C over the course of 24 hours.

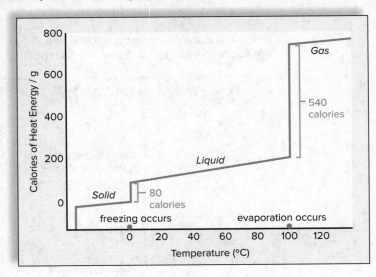

_____

_____

EK 2.A.3
Student Edition p. 28

Draw a simple diagram of how water travels through a tree. **Label** where adhesion and cohesion occurs.

EK 2.A.3
Student Edition p. 28

Describe how water can be used as a solvent to wash road salt, $CaCl_2$, off of a car.

_____

_____

SUMMARIZE IT

EK 2.A.3

What is hydrogen bonding?

_____

_____

Section 2.3 Chemistry of Water (continued)

EK 2.A.3a

Where can we find hydrogen bonds in nature?

_____

_____

EK 2.A.3a1

Why is hydrogen bonding one of the most important bonds, even though it's considered a 'weak' bond?

_____

_____

_____

EK 2.A.3

How does hydrogen bonding directly impact an organism's ability to grow, reproduce, or maintain organization?

_____

_____

_____

Section 2.4 Acids and Bases, *pp. 30–32*

EXTENDING KNOWLEDGE

This section takes the AP Essential Knowledge you have learned further, and may provide illustrative examples useful for the AP Exam.

REVIEW IT

Student Edition p. 30

Describe what makes a chemical an acid or a base.

_____

_____

Student Edition p. 30

Identify if these solutions have more hydrogen ions, more hydroxide ions, or an equal number of hydrogen and hydroxide ions based on their pH.

Solution	pH	Ion
Coffee	5	
Tears	7.0	
Soda	3.0	
Bleach	12.0	

Section 2.4 Acids and Bases (continued)

Student Edition p. 31

Give two reasons why we use the **pH scale**.

1. _____

2. _____

USE IT

Student Edition p. 30

Classify these solutions as acidic (A), basic (B), or neutral (N).

SOLUTION	pH	A, B, or N	SOLUTION	pH	A, B, or N
Orange Juice	3.5	_____	Tidal Marsh Water	8.3	_____
Human Blood	7.3	_____	Clog Remover	12.5	_____
Coconut Milk	7.0	_____	Rainwater	5.4	_____

Student Edition p. 31

Explain what **buffers** are and how human blood is buffered to maintain a neutral pH.

_____

_____

_____

SUMMARIZE IT

Review the Impact of Acid Deposition on page 31. With the decrease in chemicals such as SO_2 and NO_x in fossil fuels, what changes in pH would you hypothesize will occur both in rainwater and in lakes and streams?

_____

_____

_____

These questions were posed in the *Biology* chapter opener (page 19). Answer them using the knowledge you've gained from this chapter.

1. Why do living organisms require matter and free energy, and from where do they get it?

   _____

   _____

   _____

2. How do subatomic particles determine the chemical properties of an atom and its bonding tendencies?

   _____

   _____

   _____

3. How are water's unique properties important to life on Earth?

   _____

   _____

   _____

   _____

3 The Chemistry of Organic Molecules

As you work through your AP Focus Review Guide, keep this chapter's Big Ideas in mind:

FOLLOWING THE BIG IDEAS

 BIG IDEA 1 The diversity of biological life is the result of changes in DNA sequences and the biomolecules for which they code.

 BIG IDEA 2 Carbon's stable and versatile covalent bonding leads to tremendous variety in the carbohydrates, lipids, proteins, and nucleic acids observed in living organisms.

 BIG IDEA 3 Nucleic acids are unique among the biomolecules in their ability to store and transmit genetic information.

 BIG IDEA 4 Macromolecules form the functional basis for all cellular systems.

Section 3.1   Organic Molecules, *pp. 36–38*

Essential Knowledge Covered:
4.A.1

REVIEW IT

Given the definition on the left, **fill in** the correct vocabulary word on the right.

Vocabulary

dehydration reaction
functional group
hydrolysis (origin reaction)
inorganic chemistry
isomers
monomer
organic chemistry
organic molecules
polymers

Definition	Vocabulary Word
a specific combination of bonded atoms attached to a carbon skeleton	
organic molecules with identical molecular formulas but with different arrangements of atoms	
a chain of subunits that make up a biomolecule	
the branch of chemistry initially concerned with nonliving things	
molecules that contain carbon and hydrogen atoms	
synthesis of biomolecules through the removal of H_2O	
the branch of chemistry initially concerned with living things	
a subunit of a biomolecule	
the breakdown of biomolecules through the addition of H_2O	

USE IT

EK 4.A.1
Student Edition pp. 36–38

Complete the following chart of biomolecules, and what their monomers and polymers are called.

Biomolecule	Monomers	Polymer
	Amino acids	
Carbohydrate		Polysaccharide
		DNA, RNA
Lipids	Glycerol, fatty acids	

EK 4.A.1
Student Edition pp. 36–37

Carbon has unique properties which allows for many different types of molecules to be formed from it. **Describe** how the property allows for the formation of different biomolecules.

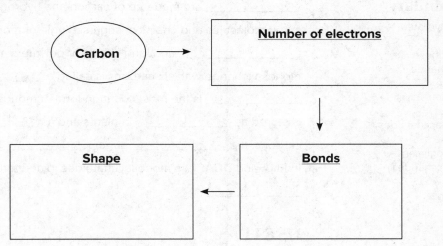

EK 4.A.1
Student Edition p. 37

Determine whether or not the following statements are true or false (T/F) about functional groups.

_____ *R* is where the functional group attaches to a carbon skeleton.

_____ A functional group determines an organic molecule's polarity.

_____ A function group does not determine the types of reactions an organic molecule will undergo.

_____ The carbon skeleton acts as a framework to position the functional group.

Section 3.1 Organic Molecules (continued)

EK 4.A.1

SUMMARIZE IT

Using Table 3.1 on page 37 of your textbook, **describe** two different functional groups. Where are they found and why are they significant? What do all functional groups have in common?

_____

_____

_____

Section 3.2 Carbohydrates, *pp. 39–42*

Essential Knowledge Covered:
4.A.1

REVIEW IT

Using the vocabulary on the left, **fill in the blanks** below.

Vocabulary

carbohydrates
chitin
deoxyribose
disaccharide
glucose
glycogen
hexoses
monosaccharaides
pentose
peptidoglycan
polysaccharides
ribose
starch

_____ are made up of carbon and hydrogen, and include both single sugar molecules and chains of sugars. Single or monomer subunits are called _____, and long chains or polymers are called _____. A monosaccharide with six carbons is called a _____. The hexose _____ is the most common form of monosaccharide cellular fuel. Glucose is stored as _____ in plants and as _____ in animals. With five carbons, the _____ sugars _____ and _____ are present in RNA and DNA. Two monosaccharaides joined through dehydration is called a _____.

EK 4.A.1
Student Edition p. 39

USE IT

What are the two major roles that carbohydrates play in living organisms?

1. _____

2. _____

EK 4.A.1
Student Edition p. 41

Name three structural polysaccharides described in this section.

1. _____

2. _____

3. _____

Section 3.2   Carbohydrates (continued)

EK 4.A.1

SUMMARIZE IT

Cellulose and starch are both made out of glucose and both are found in plants. However, humans can metabolize starch but not cellulose. What is the difference between these two carbohydrates that causes this?

_____

_____

_____

Section 3.3   Lipids, *pp. 42–46*

Essential Knowledge Covered:
4.A.1

REVIEW IT

Using the vocabulary, **identify** which lipid plays what function in the organisms listed below.

Vocabulary

fats
lipids
phospholipid
steroids
waxes

Organism	Function	Lipid
Succulent plant	Helps keep water in leaves during hot desert days	
Sunflower	Helps store energy in seeds	
Penguin	Helps bird withstand the cold Antarctic	
Peacock	Determines difference between males and females	
All organisms	Maintains structure and function of a cell	

USE IT

EK 4.A.1
Student Edition p. 43

Use the chart below to define unsaturated and saturated fatty acids.

unsaturated fatty acids.

Fatty acid

saturated fatty acids.

Section 3.3 Lipids (continued)

EK 4.A.1
Student Edition p. 45

The phospholipid bilayer plays a major role in the structure and function of a cell. **Finish** the diagram of a plasma membrane below by adding a phospholipid bilayer and **label** the **polar heads** and **nonpolar tails.**

Outside cell

Inside cell

Describe why the phospholipids arrange in the above orientation.

_____

_____

SUMMARIZE IT

EK 4.A.1

Circle all the triglycerides listed below.

Phospholipids	**Glycerol**	**Unsaturated fatty acids**
Saturated fatty acids	**Wax**	**Steroids**

EK 4.A.1

How are the lipids you didn't circle different in structure from triglycerides?

_____

_____

_____

Section 3.4   Proteins, *pp. 46—50*

Essential Knowledge Covered:
3.A.1, 4.A.1

EK 4.A.1
Student Edition p. 46

USE IT

Describe three functions that proteins play in the cell of a living organism.

_____

_____

_____

EK 4.A.1
Student Edition p. 47

Glutamine is one of the 20 amino acids that make up polypeptides.

glutamine (Gln)

Using arrows, **identify** the **amino group**, the **acidic group**, and the **R group**.

Is the R group nonpolar, polar, or ionized?

EK 3.A.1
Student Edition p. 49

Identify the level of protein organization given its definition.

Definition	Structure
The folding that results in the final-three dimensional shape of a polypeptide	
The linear sequence of amino acids encoded for by DNA	
The coiling or folding (such as α helices or β sheets) of a polypeptide	
The interaction of two or more folded polypeptides.	

SUMMARIZE IT

EK 3.A.1

With a small drawing, **show** how a dipeptide bond is synthesized from two amino acids. Be sure to include the water molecule that is released and identify the reaction that occurs to make this happen.

Section 3.5   Nucleic Acids, *pp. 50–52*

Essential Knowledge Covered:
4.A.1

REVIEW IT

Which nucleotides store the instructions for life?

_____

What are nucleotides called that store a large amount of energy used in synthetic reactions?

_____

USE IT

EK 4.A.1
Student Edition p. 51

Identify the three molecules that make up a nucleotide, using the following illustration for guidance.

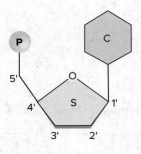

_____

_____

_____

EK 4.A.1
Student Edition p. 51

Write in the missing complementary base pairs for the following DNA sequence.

| 5' | T | T | A | ____ | ____ | 3' |
| 3' | ____ | ____ | ____ | G | C | 5' |

If this strand was RNA, how would it be different?

_____

Identify which bases are pyrimidine and which are purine, and how they differ in structure.

_____

_____

What type of bond holds complementarily base pairs together?

_____

SUMMARIZE IT

EK 4.A.1

Name two functions which DNA has in a cell.

_____

_____

EK 4.A.1

What are three types of RNA and what do they do in a cell?

_____

_____

_____

EK 4.A.1

Why is ATP called the energy currency of cells?

_____

_____

These questions were posed in the *Biology* chapter opener (page 35). Answer them using the knowledge you've gained from this chapter.

1. What elements comprise each of the four groups of macromolecules?

   _____

   _____

   _____

2. How are complex polymers synthesized from simpler monomers?

   _____

   _____

   _____

3. How does the molecular structure of the four groups of macromolecules determine the function(s) of each group?

   _____

   _____

   _____

4. How can changes in environmental conditions change the function of a macromolecule?

   _____

   _____

   _____

4 Cell Structure and Function

As you work through your AP Focus Review Guide, keep this chapter's Big Ideas in mind:

FOLLOWING THE BIG IDEAS

BIG IDEA 1 All cells are produced from existing cells, creating an unbroken lineage back to the first cells almost four billion years ago.

BIG IDEA 2 Eukaryotic cells contain multiple cooperating and specialized organelles which produce the structure and accomplish the functions necessary for life.

BIG IDEA 3 Every cell contains DNA (with or without a nuclear cover) which encrypts the information for all of its structures and functional molecules.

BIG IDEA 4 Cellular systems metabolize and adapt to changing environmental conditions.

Section 4.1   Cellular Level of Organization, *pp. 58–61*

Essential Knowledge Covered:
2.A.3, 4.A.3

Vocabulary

cell
cell theory
microscopy
surface-area-to-volume-ratio

REVIEW IT

Review Figure 4.2 on page 59 of your textbook to describe the size of life and its living components and what instruments we need to observe each (E = eye, LM = light microscope, EM = electron microscope).

Living organism or competent	Approximate Size (metric unit)	Instrument Needed to Observe
Human		
Human Egg		
Human Cell		
Bacterial Cell		
Virus		
Protein		
Amino Acid		
Atom		

Name two scientists who played a role in developing the cell theory.

_____

Section 4.1 Cellular Level of Organization (continued)

EK 4.A.3
Student Edition p. 61

USE IT

Using the vocabulary and your knowledge of cells, **fill in the blanks** to review the three postulates of the **Cell Theory.**

The Cell Theory

1. All organisms are composed of _____.

2. Cells are the basic units of _____ in organisms.

3. Cells only come from _____ cells because cells are _____.

SUMMARIZE IT

EK 2.A.3

Which figure has more surface area per volume?

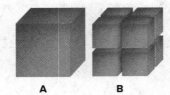

A   B

_____

If these cubes were cells, which would be better at exchanging molecules, such as nutrients or waste with the outside environment and why?

_____

Section 4.2 Prokaryotic Cells *pp. 62–63*

Essential Knowledge Covered:
2.B.3, 4.A.2

REVIEW IT

In this section, you were introduced to three different types of cells: prokaryotic, eukaryotic, and archaens.

What makes them similar? What makes them different?

Cell Type	Shared Trait	Unique Trait
Prokaryote		
Eukaryote		
Archaeans		

Section 4.2 Prokaryotic Cells (continued)

This chapter has *a lot* of new words. Let's start by reviewing some of the most important ones that belong to prokaryotes.

Vocabulary

archaeans
bacillus
capsule
cell envelope
cell wall
coccus
conjugation pili
cyanobacteria
cytoplasm
eukaryotic cells
fimbriae
flagella
glycocalyx
mesosomes
nucleoid
plasma membrane
plasmids
prokaryotic cells
ribosomes
spirilla
spirochete
thylakoids
vector

Prokaryotes have three basic shapes: _____, _____, and

_____ **or** _____. Cells are enclosed by a _____ which is

made up of the _____, the _____, and the _____.

Internal pouches which increase surface area for the attachement of enzymes are

called _____. A _____ is a well-organized layer of polysaccharides

that is not easily washed off.

Inside, the cell is filled with a semifluid solution of water and other molecules called

_____. A region known as the _____ hosts DNA. DNA can also be

found in circular pieces called _____. When plasmids are used for scientific

purposed to transport DNA to a different organism, this becomes known as a

_____. Proteins are synthesized by tiny structures called _____. In

photosynthetic bacteria or _____ their cytoplasm contains _____

where pigments such as chlorophyll produce carbohydrates.

Outside the cell, prokaryotes can have appendages or _____ which allow

them to move, _____ which allow them to stick to things, or _____

which allow for the movement of DNA to be passed from cell to cell.

USE IT

EK 2.B.3
Student Edition p. 64

Draw a diagram of a bacillus prokaryote. Include and identify the following structures on your diagram.

Structure
Flagellum
Nucleoid
Plasma Membrane
Ribosome
Cell Wall
Fimbriae
Mesosome

SUMMARIZE IT

EK 4.A.2

Compare and contrast prokaryotic and eukaryotic ribosomes.

_____

_____

EK 2.B.3

Describe the role of thylakoids in cyanobacteria.

_____

_____

Section 4.2 Prokaryotic Cells (continued)

EK 2.B.3

Why do scientists think that ancestral cyanobacteria were the earliest photosynthesizers on Earth?

_____

_____

Section 4.3 Introduction to Eukaryotic Cells, *pp. 64–67*

Essential Knowledge Covered:
1.B.1, 2.B.3, 4.A.2, 4.B.2

REVIEW IT

Identify the three major components of a eukaryotic cell.

Vocabulary

endosymbiotic theory
cytoskeleton
organelles
vesicles

Definition	Structure
An extensive network of protein fibers which maintains cell shape and assists with cell movement	
The compartments of a eukaryotic cell that carry out specialized functions	
Membranes sacs that enclose molecules and are transported around the cell	

USE IT

EK 4.B.2
Student Edition pp. 66–67

Describe how the cytoskeleton is used to send a message from the endoplasmic reticulum to the Golgi apparatus.

_____

_____

_____

EK 4.A.1
Student Edition pp. 64–66

Name one of the major organelles found in plants but not animal cells. What function does this organelle give plants that animals lack?

_____

_____

EK 2.B.3
Student Edition p. 64

Describe two reasons why internal membranes are so important in eukaryotic cells.

1. _____

_____

2. _____

_____

Section 4.3 Introduction to Eukaryotic Cells (continued)

EK 1.B.1

SUMMARIZE IT

The endosymbiotic theory was developed by Dr. Lynn Margulis. In her 1967 paper, Dr. Margulis described how eukaryotic cells arose from prokaryotic cells. How do the internal membranes around organelles support her theory?

Using this simple diagram, **label** and **describe** how organelles are thought to have evolved.

Prokaryotic cells   Eukaryotic Cell

_____

_____

_____

Section 4.4 The Nucleus and Ribosomes, *pp. 67—69*

Essential Knowledge Covered:
2.B.3, 4.A.2, 4.B.2

REVIEW IT

Given the definition on the left, **fill in** the correct structure on the right.

Vocabulary

chromatin
chromosomes
genes
nuclear envelope
nuclear pore
nucleolus
nucleoplasm
signal peptide

Definition	Structure
The command center of the cell	
The area where ribosomal RNA is produced	
A double membrane around the nucleus	
Where protein synthesis occurs	
Passages where proteins move to and from the nucleus to the cytoplasm	
Rod-like structures made of coiled chromatin	
Units of heredity	
A network of strands in the nucleoplasm which contains DNA, protein, and some RNA	

Section 4.4 The Nucleus and Ribosomes (continued)

EK 2.B.3
Student Edition pp. 67–68

Identify the following structures on the nucleus illustrated below.

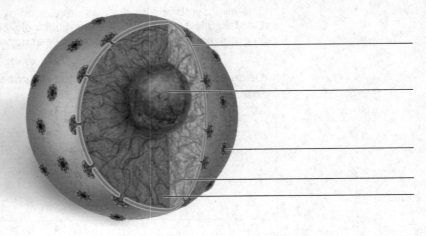

USE IT

EK 4.B.2
Student Edition pp. 67–68

Why is the nucleus an important to cell structure and function?

_____

_____

EK 4.A.2
Student Edition p. 68

How does a message intended for another part of the cell get in and out of the nuclear envelope?

_____

_____

SUMMARIZE IT

EK 4.A.2

What is the central dogma of molecular biology, and how does it govern all living cells?

_____

_____

EK 4.A.1

Illustrate how a ribosome synthesizing a protein with a signal peptide moves from the nucleus to the endoplasmic reticulum.

Section 4.5   The Endomembrane Structure, *pp. 69–71*

Essential Knowledge Covered:
1.B.1, 2.B.3, 3.A.3, 4.A.2, 4.B.2

Vocabulary

endomembrane
endoplasmic reticulum
Golgi apparatus
lysosomes
rough ER
secretion
smooth ER

REVIEW IT

There is a lot going on in a cell!
Let's review the endomembrane system.

Identify the structures that are missing from the endomembrane system on the following list.

1. The nuclear envelope

2.

3.

4. Vesicles

Fill out the missing the structure or function on the chart below.

The Endomembrane System	
Structure	**Function**
	a system of membrananous channels and saccules
The Golgi appartus	
	membrane-bound vesicles produced by the Golgi apparatus that aid in digesting cellular material
	a double membrane around the nucleus that seperates it from the cytoplasm

USE IT

EK 1.B.1, 4.A.2
Student Edition p. 69

Describe two roles the endomembrane systems serves.

1. _____

2. _____

Section 4.5   The Endomembrane Structure (continued)

EK 2.B.3
Student Edition p. 69

Compare and contrast the smooth endoplasmic reticulum with the rough endoplasmic reticulum.

Smooth ER	Rough ER
Both ERs	

EK 3.B.3
Student Edition p. 70

A rare inherited disorder called Tay-Sachs diseases is characterized by a progressive degeneration of the nervous system due to a missing lysosomal enzyme. How are nerve cells affected by the loss of this enzyme?

_____

_____

SUMMARIZE IT

EK 1.B.1, 4.A.2

Illustrate and describe how a transport vesicle enters the cis face of the Golgi apparatus and exits through the trans face to secrete its contents outside of the cell. Be sure to identify the structures in which the vesicle leaves and arrives at.

Section 4.6   Microbodies and Vacuoles, *pp. 72–73*

Essential Knowledge Covered:
2.B.3, 4.A.2, 4.B.2

Vocabulary

central vacuole
peroxisomes
vacuoles

REVIEW IT

Name the two organelles that do not communicate with the endomembrane system.

1. _____

2. _____

Identify which organelle is which.

These membranous sacs called, _____, are used for storing and breaking down waste.

All _____ contain enzymes whose reaction result in the production of hydrogen peroxide.

Section 4.6   Microbodies and Vacuoles (continued)

EK 4.A.2
Student Edition p. 75

USE IT

How does the central vacuole provide structural support to the cell?

_____

EK 2.B.3
Student Edition p. 76

In what way is the central vacuole in plants analogous to the lysosomes in animal cells?

_____

_____

SUMMARIZE IT

EK 4.B.2
Student Edition pp. 75—76

Compare and **contrast** a peroxisome and a vacuole.

_____

_____

_____

Section 4.7   The Energy—Related Organelles, *pp. 73—75*

Essential Knowledge Covered:
1.B.1, 2.B.3, 4.A.2, 4.B.2

REVIEW IT

Life depends on a constant input of energy to maintain the structure of cell.

Vocabulary

chloroplasts
chromoplasts
cristae
leuoplasts
granum
matrix
mitochondria
plastids
stroma
thylakoids

Identify the two organelles specialize in converting energy to a usable form.

1. _____

2. _____

Define what a plastid is, and list three that can be found in plant cells.

_____

_____

_____

USE IT

EK 4.A.2
Student Edition pp. 73—74

Identify which organelle can be represented by which equation.

carbohydrates + oxygen → carbon dioxide + water + ATP

_____

solar energy + carbon dioxide + water → carbohydrate + oxygen

_____

Section 4.7 The Energy–Related Organelles (continued)

EK 1.B.1
Student Edition pp. 76—77

Describe two important features about chloroplasts and the mitochondria which support the endosymbiotic theory.

_____

_____

_____

EK 2.B.3
Student Edition p. 77

Chloroplasts have a third membrane called the thylakoids. **Describe** what role they play.

_____

_____

EK 2.B.3
Student Edition p. 77

Mitochondria have two membranes. How is the inner membrane different than the outer membrane?

_____

_____

_____

SUMMARIZE IT

EK 4.B.2

While only plants and algae contain chloroplasts, all eukaryotes contain mitochondria which produce ATP. Why is ATP important?

_____

_____

_____

Section 4.8   The Cytoskeleton, *pp. 75—77*

Essential Knowledge Covered:
1.B.1, 2.B.3, 4.A.2, 4.B.2

REVIEW IT

Use the vocabulary words on the left to review the important structures in the cytoskeleton.

Vocabulary

actin filaments
basal body
centrioles
centrosome
flagella
intermediate filaments
microtubules
motor molecules

The cytoskeleton is a dynamic network of proteins, giving the cell shape and allowing

the cell and its organelles to move. Proteins which attach, detach, and reattach to

filaments in the cell are called _____. There are three types of filaments.

_____ are long, very thin, and made up two chains of twisted globular

actin monomers. Between the size of actin filaments and microtubules, are the

_____ which form ropelike assemblies of polypeptides. _____

are made of globular tubulin and controlled through the _____. A short cylinder

of microtubules arranged in an outer ring is called a _____. The organelle called

the _____ may direct the organization of microtubules within _____

and _____, two hair-like projections which may aid in their movement.

Section 4.8   The Cytoskeleton (continued)

USE IT

Fill in the chart below.

Filament Name	Protein subunit	Associated motor molecule
actin		
	fibrous subunit (such as kertin)	n/a
		kinesin, dynein

EK 2.B.3, EK 4.A.2
Student Edition p. 75

Compare and contrast the function of actin filaments and microtubules in a cell.

_____

_____

EK 1.B.1
Student Edition p. 76

In your own words, how does a chameleon use the cytoskeleton to change the color of its skin to blend into its environment?

_____

_____

_____

EK 1.B.1
Student Edition p. 77

Describe how cilia and flagella are able to move using the cytoskeleton.

_____

_____

EK 4.A.2
Student Edition p. 75

List two function intermediate filaments can have in a cell.

_____

_____

_____

Section 4.8   The Cytoskeleton (continued)

SUMMARIZE IT

Choose one of the following two diagrams to answer the questions below.

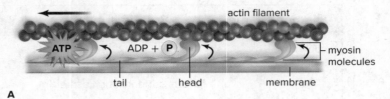

A

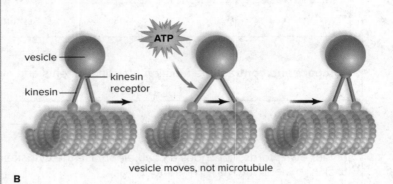

vesicle moves, not microtubule

B

EK 1.B.1
Student Edition pp. 75–76

Identify the filament and the motor protein.

_____

Describe the interaction that is occurring.

_____

_____

EK 4.B.2
Student Edition pp. 75–76

What process in a living organism may be the result of this interaction?

_____

_____

_____

What role does ATP play in the process?

_____

_____

These questions were posed in the *Biology* chapter opener (page 57). Answer them using the knowledge you've gained from this chapter.

1. Why are most cells so small?

_____

_____

_____

2. What structures do all cells share? What evidence supports the theory that eukaryotes evolved from prokaryotic cells?

_____

_____

_____

3. What are structural differences between the genetic material of prokaryotes and eukaryotes?

_____

_____

_____

4. What features allow eukaryotic cells to function?

_____

_____

_____

5 Membrane Structure and Function

As you work through your AP Focus Review Guide, keep this chapter's Big Ideas in mind:

FOLLOWING THE BIG IDEAS

BIG IDEA 2 The plasma membrane, a feature of all cells, is appropriately called the gatekeeper of the cell because it maintains the identity and integrity of the cells as it "stands guard" over what enters and leaves.

BIG IDEA 3 Membrane receptor proteins act as intercellular signal receivers.

BIG IDEA 4 Membranes are an integral part of an interconnected cellular system of communication and response to environment.

Section 5.1   Plasma Membrane Structure and Function, *pp. 83–87*

Essential Knowledge Covered:
2.B.2, 3.A.1, 3.B.2, 3.D.1,
3.D.2, 4.C.1

REVIEW IT

Given the definition on the left, **fill in** the correct vocabulary word on the right.

Vocabulary

aquaporins
bulk transport
concentration gradient
extracellular matrix
glycolipid
glyoprotein
selectively permeable

Definition	Vocabulary Word
Only allows certain substances in, while keeping others out	
Channel proteins that allow water across a membrane	
Where molecules flow from where their concentration is high to where their concentration is low	
A phospholipid attached to a carbohydrate chain	
A way in which large particles can exit or enter a cell	
Protein and carbohydrate molecules found outside of animal cells	
A protein with an attached carbohydrate chain	

USE IT

Describe what causes cystic fibrosis and how it affects the body.

EK 3.A.1
Student Edition p. 85

_____

_____

EK 3.B.2, 4.C.1
Student Edition p. 85

Identify each membrane protein and describe its function.

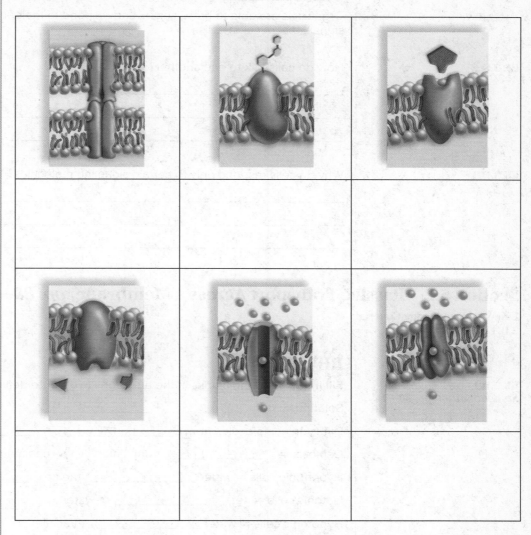

Carbon dioxide is an example of a small, noncharged molecule that can pass across a cell's membrane.

EK 2.B.2
Student Edition p. 86

Which way will the CO_2 move through this membrane?

CO_2 CO_2 CO_2 CO_2 CO_2 CO_2		CO_2 CO_2 CO_2
Outside cell	**Cell membrane**	**Inside cell**

Section 5.1 Plasma Membrane Structure and Function (continued)

SUMMARIZE IT

Organisms are able to grow, sense, and respond to their environment due to specific signals that occur in cells.

EK 3.D.2

How do cells talk to one another?

_____

_____

_____

EK 3.D.1

Why do cells respond only to certain signaling molecules?

_____

_____

_____

Section 5.2 Passive Transport Across a Membrane, *pp. 88–91*

Essential Knowledge Covered:
2.B.1, 2.B.2

REVIEW IT

EK 2.B.1
Student Edition pp. 88–91

Fill in the following blanks in the equations in order to define each vocabulary word.

Solution = _____ + _____

Diffusion = molecules from a _____ → _____ concentration

Osmosis = _____ from a high → low concentration

Hypertonic cells = water _____ the cell

Isotonic cells ≠ net _____ of water

Hypotonic cells = water _____ the cell

USE IT

EK 2.B.2
Student Edition p. 90

A laboratory technician performed a routine red blood cell count. The cells appeared shriveled and were difficult to count. What was wrong with the solution the cells were stored in?

_____

_____

EK 2.B.1
Student Edition p. 90

The technician from the question above then noticed that the cells were not in the correct solution and diluted the solution to a concentration of 0.5% sodium chloride. What happened to the cells?

_____

_____

Section 5.2   Passive Transport Across a Membrane (continued)

EK 2.B.2
Student Edition pp. 90—91

Imagine that you find a wilted piece of celery lurking in your refrigerator. You take it out and put it in a glass of water. The next day, the celery stalk is firm again. Explain what has happened at the cellular level. Use the words *vacuole, osmotic pressure, turgor pressure, hypertonic,* and *hypotonic* in your answer.

_____

_____

_____

_____

EK 2.B.2
Student Edition pp. 90—91

Compare and contrast how organisms have evolved to live in fresh water versus how organisms have evolved to live in salt water.

_____

_____

_____

_____

SUMMARIZE IT

EK 2.B.1, 2.B.2

Compare and contrast diffusion and facilitated transport.

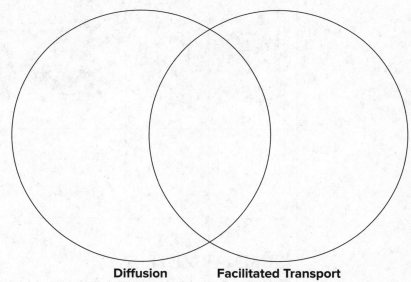

Diffusion   **Facilitated Transport**

Section 5.3   Active Transport Across a Membrane, *pp. 91—94*

Essential Knowledge Covered:
2.B.2

Vocabulary

active transport
endocytosis
exocytosis
phagocytosis
pinocytosis
receptor-mediated
endocytosis
sodium-potassium pump

EK 2.B.2
Student Edition pp. 92—94

REVIEW IT

Using the vocabulary on the left, **fill in the blanks** to review the Learning Outcomes of this section.

_____ requires energy in order to transport a molecule against its concentration gradient. The _____ is a carrier protein that binds three sodium ions and changes shape as it moves sodium across the membrane to pick up two potassium ions. Large molecules may exit a cell through _____ and enter a cell through _____. When a cell engulfs debris through endocytosis, this process is called _____. _____ occurs when vesicles form around liquid or tiny particles. A highly selective form of pinocytosis is called _____ endocytosis.

USE IT

Identify the form of endocytosis below and describe its process.

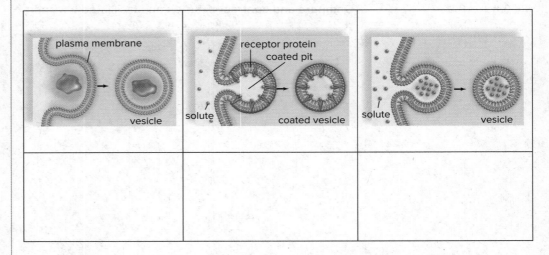

SUMMARIZE IT

EK 2.B.2
Student Edition p. 93

Design a flow chart to illustrate how a cell secretes substances outside of the cell through exocytosis. Be sure to label the plasma membrane, the secretory vesicle, and the substance being transferred.

Section 5.4 Modification of Cell Surfaces, *pp. 95–96*

Essential Knowledge Covered:
2.B.1, 2.C.2, 3.D.2

EK. 2.B.1
Student Edition pp. 95–96

REVIEW IT

Identify the two types of animal cell surface features described in this section.

1. _____

2. _____

EK 2.C.2
Student Edition pp. 95–96

There are many components in the extracellular matrix. This section briefly covered a few. **Review** what they were and how they function.

The Extracellular Matrix	
Components	**Function**
Collagen and Elastin	
	Resists compression and assist in cell signaling
	Influences the shape and activity of the cell by allowing for communication between the EMC and cytoskeleton

Name the three junctions between cells discussed in this section.

_____

USE IT

All plants have cell walls.

EK 2.B.1, 3.D.2
Student Edition p. 96

Describe the importance of these three components to a cell wall and where they are found.

Pectin	Plasmodesmata	Lignin

Section 5.4   Modification of Cell Surfaces (continued)

EK 2.B.1
Student Edition pp. 95–96

Place an X under the junction where the following statement applies.

	Desmosomes	Tight Junction	Gap Junction
This junction allows communication between two cells by joining plasma membrane channels			
This junction forms an impermeable barrier between adjacent cells			
Mechanically attaches to adjacent cells			

SUMMARIZE IT

EK 2.B.1

Identify the junction that helps play a role in organismal processing.

Process	Junction
Keeps digestive juice only moving between intestinal cells	
Passes water from plant cell to plant cell	
Allows for a flow of ions to between heart cells, permitting contractions	
Holds skin cells together and allows skin to remain elastic	

EK 2.C.2

Why might plants have evolved cell walls while animals did not?

_____

_____

_____

These questions were posed in the *Biology* chapter opener (page 82). Answer them using the knowledge you've gained from this chapter.

1. How does the fluid mosaic model of the cell membrane allow for selective permeability?

   _____

   _____

   _____

2. How do signaling pathways detect and respond to changes in a cell's environment?

   _____

   _____

   _____

3. How do membrane-bound organelles in eukaryotic cells confer greater efficiency to cell processes?

   _____

   _____

   _____

6 Metabolism: Energy and Enzymes

As you work through your AP Focus Review Guide, keep this chapter's Big Ideas in mind:

FOLLOWING THE BIG IDEAS

 BIG IDEA 1   Cells have evolved to metabolize energy in order to support cellular processes important to life.

 BIG IDEA 2   Understanding the principles of metabolism and energy transformation and transfer helps us understand how cells and organisms function.

 BIG IDEA 4   Energy flows through biological systems, resulting in the ability to do work, even though with each transfer energy is lost as heat.

Section 6.1   Cells and the Flow of Energy, *pp. 101–102*

Essential Knowledge Covered:
2.A.1, 2.A.2

EK 2.A.1
Student Edition p. 101

EK 2.A.1
Student Edition p. 101

EK 2.A.1
Student Edition pp. 101–102

EK 2.A.1
Student Edition p. 102

REVIEW IT

Define energy.

_____

List the two forms in which energy occurs.

1. _____

2. _____

Fill in the blanks to complete the two laws of thermodynamics.

1. _____ cannot be created or destroyed, but it can be changed from one

    form to another.

2. Energy cannot be changed from one form to another without a loss of usable

    _____.

Define entropy.

_____

_____

Section 6.1 Cells and the Flow of Energy (continued)

USE IT

EK 2.A.1
Student Edition p. 101

Identify the form of energy in each description as either kinetic (K) or potential (P).

Energy Form	Description
	the monkey swinging through a tree
	the fig that the monkey eats
	molecules in the fig that the monkey eats
	the monkey eating a fig

EK 2.A.2
Student Edition p. 101

Describe what happens to the solar energy which shines down on a plant.

_____

_____

_____

EK 2.A.1
Student Edition p. 102

Use the second law of thermodynamics to describe why and how glucose breakdown over time.

_____

_____

SUMMARIZE IT

EK 2.A.1, 2.A.2

Compare and contrast chemical energy and mechanical energy.

_____

_____

_____

_____

EK 2.A.1, 2.A.2

The human body produces a lot of heat. **Describe** this heat in terms of solar energy and entropy.

_____

_____

_____

_____

_____

Section 6.2   Metabolic Reactions and Energy Transformations, *pp. 103–104*

Essential Knowledge Covered:
2.A.1, 2.A.2

Vocabulary

ADP
ATP
free energy
endergonic
exergonic
metabolism
mole

EK 2.A.2
Student Edition p. 103

REVIEW IT

Using the vocabulary on the left, **fill in the blanks** to review the Learning Outcomes of this section.

The sum of all chemical reactions that occur in a cell is called _____. The amount of energy left to do work after a chemical reaction has occurred is

_____. Spontaneous reactions are called _____, while reactions that require an input of energy are called _____. The most common energy currency of cells is _____. ATP can be regenerated from _____ and inorganic phosphate. The hydrolysis of ATP to ADP and inorganic phosphate can be measured in kcal per _____.

Identify the reactants and products in this equation.

$$A + B \rightarrow C + D$$

_____

USE IT

EK 2.A.2
Student Edition p. 103

Illustrate the ATP cycle in terms of the creation and hydrolysis of ATP. Identify which reaction is endergonic and which is exergonic.

EK 2.A.1
Student Edition p.104

List three ways which ATP provides energy for living organisms.

1.

2.

3.

EK 2.A.1, 2.A.2
Student Edition p.104

What does it mean when a product has been *phosphorylated*? You may find it helpful to draw a picture of the coupling of ATP to an energy-requiring reaction in order to help you answer this question.

Section 6.2   Metabolic Reactions and Energy Transformations (continued)

SUMMARIZE IT

EK 2.A.2

In your own words, **describe** how ATP is used to contract a muscle which is illustrated in the figure below.

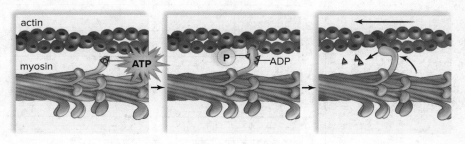

_____

_____

_____

_____

Section 6.3   Metabolic Pathways and Enzymes, *pp.* 105–109

Essential Knowledge Covered:
2.A.1, 2.D.1, 3.A.1, 4.A.2, 4.B.1

REVIEW IT

Insert the correct definition or vocabulary word in the list below.

Vocabulary

active site
co-factors/co-enzymes
denaturation
enzyme
metabolic pathway
ribozymes
substrates
vitamins

Definition	Vocabulary Word
a series of linked reactions	
	active site
a protein molecule that speeds up a chemical reaction	
	ribozymes
a frequent component of a coenzyme often found in our diets	
	co-factors/co-enzymes
reactants in an enzymatic reactions	
	denaturation

Section 6.3   Metabolic Pathways and Enzymes (continued)

EK 4.B.1
Student Edition pp. 106–107

List four factors that can impact enzymatic speed.

1. _____

2. _____

3. _____

4. _____

EK 4.B.1
Student Edition p. 106

Using the enzyme and substrate below, **describe** the induced fit model.

_____

_____

_____

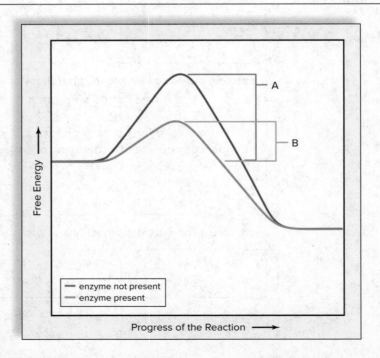

USE IT

EK 2.A.1
Student Edition p. 106

The diagram below illustrates a simple metabolic pathway, where E stands for the enzyme needed in order to complete a reaction:

$$A \quad \overset{E_1}{\rightarrow} \quad B \quad \overset{E_2}{\rightarrow} \quad C$$

If E_1 became denatured, what would happen to the pathway?

_____

_____

Section 6.3   Metabolic Pathways and Enzymes (continued)

EK 4.B.1
Student Edition p. 106

What do **A** and **B** represent? How do they differ from one another?

_____

_____

_____

EK 4.A.2
Student Edition p. 106

If the enzyme in the reaction above needed to be in solution with a pH = 5 in order to maintain structural integrity, and the solution was at pH 8, what would happen to the reaction?

_____

_____

EK 2.D.1
Student Edition p. 106

Decide whether or not the following factors affect the rate of an enzymatic reaction. Place a + if the factor increases the rate, a − if it decreases the rate, or an = sign if the rate stays the same.

Factor	Rate of Reaction
The concentration of substrate increases	
An inhibitor is present	
Optimal pH has been obtained	
A cofactor is missing	

EK 4.A.2
Student Edition p. 107

List the three critical cofactors that play a role in either cellular respiration or photosynthesis.

_____

SUMMARIZE IT

EK 4.B.1

Compare and contrast these two different forms of enzyme inhibition: noncompetitive inhibition and competitive inhibition.

_____

_____

_____

_____

EK 3.A.1

Some animals, such as Siamese cats, have what is called point coloration where the warmest parts of the body are pale and the cooler extremities are darker in color. This is caused by a mutated gene that regulates an enzyme involved in melanin production. How might this enzyme work?

_____

_____

_____

Section 6.4 Oxidation–Reduction Reactions and Metabolism, *pp. 109–110*

Essential Knowledge Covered:
2.A.1, 2.A.2

EK 2.A.1
Student Edition p. 109

EK 2.A.2
Student Edition p. 109

EK 2.A.2
Student Edition p. 109

EK 2.A.2
Student Edition p. 110

EK 2.A.2

REVIEW IT

This section introduces a very important theme!

Describe what happens in a redox reaction (think of the term OIL RIG).

_____

_____

List the two important metabolic pathways to all of life that contain redox reactions.

1. _____

2. _____

Name the two organelles involved in a redox cycle.

_____

USE IT

Satisfy the following equations.

A. energy + _____ CO_2 + 6_____ → _____ + 6 O_2

B. $C_6H_{12}O_6$ + 6 _____ → _____ + 6 CO_2 + $6H_2O$

Identify which equation describes cellular respiration and which describes photosynthesis.

_____

Which molecules are reduced and which are oxidized in photosynthesis?

_____

Which molecules are reduced and which are oxidized in cellular respiration?

_____

SUMMARIZE IT

Illustrate the redox cycle that occurs between a mitochondria and a chloroplast. Be sure to include solar energy, ATP, heat, and the molecules involved in the redox reaction.

AP Reviewing the Essential Questions

These questions were posed in the *Biology* chapter opener (page 100). Answer them using the knowledge you've gained from this chapter.

1. How does the structure of ATP enable the molecule to power cellular work?

   _____

   _____

   _____

2. How do the first and second laws of thermodynamics relate to cell metabolism?

   _____

   _____

   _____

3. How do enzymes facilitate the chemical reactions that constitute metabolism?

   _____

   _____

   _____

4. How can changes in environmental conditions and other factors affect the rate of an enzyme-catalyzed reaction?

   _____

   _____

   _____

7 Photosynthesis

As you work through your AP Focus Review Guide, keep this chapter's Big Ideas in mind:

FOLLOWING THE BIG IDEAS

 BIG IDEA 1 Plants have evolved the ability to capture solar energy and store it in carbon-based organic nutrients.

 BIG IDEA 2 All life on Earth depends on the energy stored in carbohydrates produced through photosynthesis.

 BIG IDEA 4 Organic nutrients produced by plants through photosynthesis are passed on to other organisms in a food web that have evolved to feed on plants, thus transferring free energy among members of the web.

Section 7.1 Photosynthetic Organisms, *pp. 115—116*

Essential Knowledge Covered:
2.A.2

EK 2.A.2
Student Edition p. 116

REVIEW IT

Match the plant component to its definition.

chlorophyll	Small openings on the leaf where carbon dioxide enters
thylakoid	Pigment which absorbs solar energy and drives photosynthesis
stomata	A double membrane surrounding the chloroplast
stroma	Stacks of thylakoids
chloroplast	A membrane within the stroma which forms flattened stacks
grana	Organelles which carry on photosynthesis

EK 2.A.2
Student Edition p. 115

USE IT

Identify if the organism is a heterotroph (H) or an autotroph (A).

_____ worm

_____ plant

_____ mammal

_____ cyanobacteria

_____ algae

Section 7.1   Photosynthetic Organisms (continued)

EK 2.A.2

SUMMARIZE IT

Compare and contrast a heterotroph and an autotroph.

_____

_____

_____

Section 7.2   The Process of Photosynthesis, *pp. 117–118*

Essential Knowledge Covered:
2.A.1, 2.A.2

Vocabulary

Calvin cycle
light reaction

EK 2.A.2
Student Edition p. 117

REVIEW IT

Reduction and oxidation take place in photosynthesis.

Label the diagram below to show which molecules are reduced and which are oxidized.

$$\text{solar energy}$$
$$CO_2 + H_2O \longrightarrow (CH_2O) + O_2$$

EK 2.A.2
Student Edition pp. 117–118

Identify which equation represents the light reaction and which equation represents the Calvin cycle.

The Two Stages of Photosynthesis	
solar energy → chemical energy (ATP, NADPH)	chemical energy → chemical energy (ATP, NADPH) (carbohydrate)

USE IT

EK 2.A.2
Student Edition p. 117

Explain the role of $NADP^+$/NADPH in photosynthesis.

_____

_____

_____

Section 7.2   The Process of Photosynthesis (continued)

EK 2.A.1
Student Edition p. 118

Label the light reaction and Calvin reaction and the energy that powers the reactions.

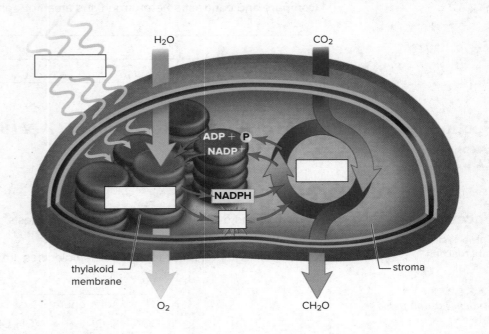

SUMMARIZE IT

EK 2.A.1, 2.A.2

Compare and contrast the light and Calvin cycle.

_____

_____

_____

_____

Section 7.3   Plants Convert Solar Energy, *pp. 119–123*

Essential Knowledge Covered:
2.A.1, 2.A.2, 4.C.1

REVIEW IT

Insert the correct definition or vocabulary word in the list below.

Vocabulary

absorption spectrum
chemiomosis
cyclic pathway
photosystem
noncyclic pathway

Definition	Vocabulary Word
Pathway in which electrons move from water to NADP⁺	
	cyclic pathway
the production of ATP by the movement of hydrogen ions	
	photosystem
Wavelengths pigments are able to absorb	

Section 7.3   Plants Convert Solar Energy (continued)

EK 2.A.2
Student Edition pp. 119–120

List four molecular complexes present in the thylakoid membrane.

_____

USE IT

EK 4.C.1
Student Edition p. 119

Why do plants appear green to us?

_____

_____

EK 4.C.1
Student Edition p. 119

When chlorophyll breaks down in leaves during the fall, why do they appear yellow and orange?

_____

_____

EK 2.A.2
Student Edition p.120

Illustrate how ATP and NADPH are used in PSII and PSI.

SUMMARIZE IT

EK 2.A.1

Where do electrons come that are used in PSII, and how are they transferred to PSI?

_____

_____

_____

Section 7.4   Plants Fix Carbon Dioxide, *pp. 123–124*

Essential Knowledge Covered:
2.A.1, 2.A.2

Vocabulary

carbon dioxide fixation
RuBP carboxylase
redox reaction

REVIEW IT

List the three steps of the Calvin cycle.

1. _____

2. _____

3. _____

USE IT

EK 2.A.2
Student Edition pp.123–124

What is the final product of the Calvin cycle and why is it important?

_____

_____

_____

Section 7.4 Plants Fix Carbon Dioxide (continued)

EK 2.A.2
Student Edition p.124

Describe two steps which make G3P. Be sure to explain how ATP and NADPH are used.

_____

_____

EK 2.A.2
Student Edition p.124

Identify two possible fates of G3P.

_____

_____

SUMMARIZE IT

EK 2.A.1

Identify the metabolites of the Calvin cycle with an arrow. **Fill in** the spaces with the correct molecules used to drive the different steps of the Calvin cycle.

Where were these molecules that provide the energy and electrons for the reduction reactions produced?

_____

How many G3P molecules are needed to form glucose?

_____

Section 7.5 Other Types of Photosynthesis, *pp. 125–126*

Essential Knowledge Covered:
2.A.1, 2.A.2

Vocabulary

C_3 plants
C_4 plants
CAM photosynthesis
photorespiration

EK 2.A.2
Student Edition p.126

EK 2.A.2
Student Edition pp.125–126

EK 2.A.2
Student Edition pp.125–126

EK 2.A.2

REVIEW IT

Using the vocabulary on the left, **fill in the blanks** to review the Learning Outcomes of this section.

The majority of plants are _____ and carry out photosynthesis using

RuBP carboxylase to fix CO_2 to RuBP. _____ occurs when stomata close

to conserve water, causing oxygen to join with RuBP instead of CO_2. In order to stop this

wasteful reaction, _____ adapted to fix CO_2 to PEP with the enzyme PEPCase.

Plants that perform _____ fix CO_2 at night and release C_4 to the Calvin

cycle during the day.

USE IT

What is the primary advantage for partitioning photosynthesis temporally in CAM plants?

_____

Determine whether or not the following statements are true or false (T/F) about the different types of photosynthesis.

_____ Plants with CAM photosynthesis are most likely found in environments where temperatures are below 25°C.

_____ C_4 plants partition the fixation of CO_2 temporally.

_____ C_3 photosynthesis partions CO_2 in mesophyll cells and the Calvin cycle in bundle sheath cells.

_____ C_4 plants partition the fixation of CO_2 in different spaces.

Complete the following chart of photosynthesis types and how they fix CO_2.

Photosynthesis Type	CO_2 fixing enzyme	Cells involved
	RuBP	
CAM		Mesophyll cells and bundle sheath cells
		Mesophyll cells

SUMMARIZE IT

Identify the following processes.

RuBP + CO_2 $\xrightarrow{\text{RuBP carboxylase}}$ 2 3PG   _____

PEP + CO_2 $\xrightarrow{\text{PEPCase}}$ oxaloacetate   _____

RuBP + O_2 $\xrightarrow{\text{RuBP carboxylase}}$ 3PG + CO_2 _____

Section 7.5   Other Types of Photosynthesis (continued)

In your own words, describe how C_4 and CAM plants are able to live in hot, dry conditions and why C_3 plants have problems.

_____

_____

_____

_____

AP Reviewing the Essential Questions

These questions were posed in the *Biology* chapter opener (page 114). Answer them using the knowledge you've gained from this chapter.

1. What types of organisms use photosynthesis to obtain free energy necessary for life processes?

_____

_____

2. How do plants capture solar energy and convert it to chemical energy of food?

_____

_____

_____

_____

3. What is the relationship between photosynthesis and cellular respiration in terms of reactants and products? How are these processes interdependent?

_____

_____

_____

_____

8 Cellular Respiration

As you work through your AP Focus Review Guide, keep this chapter's Big Ideas in mind:

FOLLOWING THE BIG IDEAS

 BIG IDEA 1 The majority of organisms on Earth use cellular respiration, indicating an ancient biological lineage.

 BIG IDEA 2 Chemical energy in the bonds of food molecules can be released in small, regulated steps through cellular respiration, transferring free energy to create ATP molecules.

 BIG IDEA 4 The energy for life typically originates with sunlight, whose solar energy passes to the chloroplast where some of it is stored in the chemical energy of carbohydrates, which are passed to mitochondria where some is stored in the chemical energy of ATP molecules.

Section 8.1   Overview of Cellular Respiration, *pp. 130–131*

Essential Knowledge Covered:
2.A.1, 2.A.2

Vocabulary

aerobic
anaerobic
cellular respiration
fermentation
glycolysis
NAD$^+$

REVIEW IT

Given the definition on the left, **fill in** the correct vocabulary word on the right.

Definition	Vocabulary Word
A coenzyme of oxidation	
The process by which cells acquire energy by breaking down nutrient molecules	
The anaerobic metabolism of pyruvate in the cytoplasm	
Phases of cellular respiration which do not require oxygen	
The breakdown of glucose	
Phases of cellular respiration which require oxygen	

EK 2.A.2
Student Edition p. 131

List the four phases of the complete breakdown of glucose.

1. _____

2. _____

3. _____

4. _____

EK 2.A.1
Student Edition p.131

USE IT

Fill in the diagram below to show the order of the four phases of cellular respiration. In each box, place the name of the phase and the main products produced. **Place** a star by the cycles that turn twice.

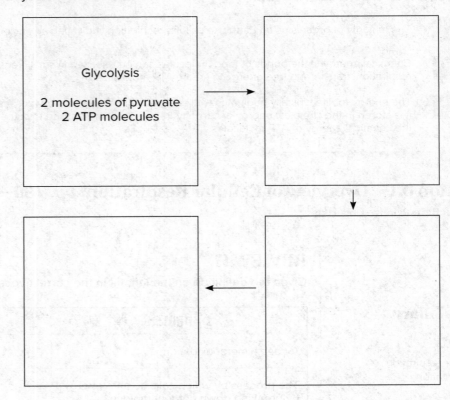

Glycolysis

2 molecules of pyruvate
2 ATP molecules

EK 2.A.1
Student Edition p.131

Identify if the phase of cellular respiration takes place inside the mitochondria or outside the mitochondria (**O** or **I**).

Circle the phases which are anaerobic.

_____ The Electron Transport Chain

_____ Glycolysis

_____ The Prep Reaction

_____ The Citric Acid Cycle

Section 8.1 Overview of Cellular Respiration (continued)

SUMMARIZE IT

EK 2.A.1

Explain why the breakdown of glucose happens in steps.

_____

_____

EK 2.A.2

Why are NAD$^+$ and FAD important in cellular respiration?

_____

_____

_____

EK 2.A.2

Describe how the human body produces ATP after a person eats a large bowl of pasta.

_____

_____

_____

_____

Section 8.2 Outside the Mitochondria: Glycolysis, *pp. 132–133*

Essential Knowledge Covered:
2.A.1, 2.A.2

REVIEW IT

EK 2.A.1
Student Edition pp. 132–133

Glycolysis is the breakdown of C$_6$ carbon _____ to two C$_3$ _____ molecules.

Where does glycolysis occur?

_____

The formation of **36 to 38 ATP** are theoretically possible when glucose is broken down completely. How many ATP molecules are produced during glycolysis?

_____

EK 2.A.1
Student Edition p.132

USE IT

Identify the inputs and outputs of glycolysis.

Glycolysis	
Inputs	**Outputs**
6C glucose	
	2 NADH
2 ATP	

SUMMARIZE IT

EK 2.A.2

Substrate-level phosphorylation occurs in the later steps of glycolysis. **Draw** a picture of how an enzyme might pass phosphate to ADP to form ATP. Be sure to label the enzyme and molecules.

Why is it thought that glycolysis takes place outside of the mitochondria?

_____

_____

_____

Section 8.3   Outside the Mitochondria: Fermentation, *pp. 134–135*

Essential Knowledge Covered:
2.A.1 , 2.A.2

EK 2.A.1
Student Edition pp. 134–135

REVIEW IT

Fermentation is the breakdown of _____ to two lactate or two alcohol and

two _____ molecules.

Where does fermentation occur?

_____

The formation of **36 to 38 ATP** are theoretically possible when glucose is broken down completely. How many ATP molecules are produced during fermentation?

_____

Identify the inputs and outputs of fermentation.

Fermentation	
Inputs	**Outputs**
glucose	2 lactate or 2 alcohol

USE IT

EK 2.A.1
Student Edition p.134

Identify the type of fermentation possible by each organism (**L** for lactate acid fermentation or **A** for those that produce ethyl alcohol as a result).

Animals ____

Plants ____

Yeast ____

Bacteria ____

Section 8.3 Outside the Mitochondria: Fermentation (continued)

SUMMARIZE IT

EK 2.A.1

Compare and contrast the two forms of fermentation.

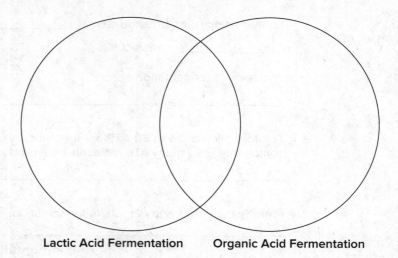

Lactic Acid Fermentation   Organic Acid Fermentation

Section 8.4 Inside the Mitochondria, *pp. 136–140*

Essential Knowledge Covered:
2.A.1, 2.A.2

REVIEW IT

EK 2.A.1, 2.A.2
Student Edition pp. 136–140

Why is the mitochondria often referred to as the powerhouse of the cell?

_____

_____

Fill in the following sentences with the correct word or number.

A maximum of _____ ATP molecules may be produced by the electron transport chain.

The _____ converts products from glycolysis into products that enter the citric acid cycle.

It takes _____ turns of the citric acid cycle to process each original glucose molecule.

Electrons that enter the electron transport chain are carried by _____.

Section 8.4 Inside the Mitochondria (continued)

USE IT

EK 2.A.2
Student Edition pp.136−138

Fill in the following information concerning the preparatory reaction, the citric acid cycle, and the electron transport chain.

	The Preparatory Reaction	The Citric Acid Cycle	The Electron Transport Chain
Products			
Location in mitochondria where the process occurs			

EK 2.A.1
Student Edition p.137

The citric acid cycle is also called the Krebs cycle. **Explain** what the cycle is responsible for during cellular respiration.

_____

_____

_____

EK 2.A.1
Student Edition p.138

Identify this process:

ADP + (P) → ATP

_____

SUMMARIZE IT

EK 2.A.2

Why is oxygen important in cellular respiration?

_____

_____

EK 2.A.2

If oxygen was not present during cellular respiration, what would happen?

_____

_____

EK 2.A.2

Cyanide is a poisonous substance because it bind to a redox carrier called cytochrome. Why would this be dangerous?

_____

_____

_____

Section 8.5   Metabolism, *pp. 141–142*

Essential Knowledge Covered:
2.A.1, 2.A.2

Vocabulary

anabolism
deamination
catabolism
metabolic pools

REVIEW IT

Using the vocabulary on the left, **fill in the blanks** to review the Learning Outcomes of this section.

Molecules needed for metabolic pathways are said to be stored in _____.

Metabolic pools are added to through the constructive reactions called _____

and broken down through _____. Proteins are sometimes also used as an

energy source in which amino acids may undergo a process called _____,

thereby losing their amino group.

Name the two organelles instrumental in allowing the flow of energy through living organism.

1. _____

2. _____

USE IT

EK 2.A.1
Student Edition p.141

Determine whether or not the following statements are true or false (T/F) about the metabolic pool.

_____ The balance of catabolism and anabolism is essential for optimum cellular function.

_____ Anabolism is the breakdown of molecules.

_____ Catabolism is the building of new molecules.

_____ Products from catabolic processes are needed in order to build new molecules.

EK 2.A.2
Student Edition p.142

Mitochondria and chloroplasts perform opposite processes but have very similar structures in their organelles. Using the table below, **describe** the differences in the organelles.

Structure	Chloroplast	Mitochondria
Inner membrane		
The Electron Transport Chain		
Enzymes		

Section 8.5 Metabolism (continued)

SUMMARIZE IT

EK 2.A.2

Why is fat an efficient form of stored energy?

_____

_____

_____

EK 2.A.2

Describe how energy flows from the Sun to a chloroplast in a plant and then to a mitochondria in a human. Feel free to illustrate your answer.

_____

_____

_____

_____

AP Reviewing the Essential Questions

These questions were posed in the *Biology* chapter opener (page 129). Answer them using the knowledge you've gained from this chapter.

1. How are the processes of photosynthesis and cellular respiration interdependent?

   _____

   _____

   _____

2. What is the role of the electron transport system in producing ATP?

   _____

   _____

   _____

3. What is the role of enzymes in regulating cellular respiration?

   _____

   _____

   _____

9 The Cell Cycle and Cellular Reproduction

As you work through your AP Focus Review Guide, keep this chapter's Big Ideas in mind:

FOLLOWING THE BIG IDEAS

 For unicellular organisms, cell division results in the formation of two new organisms, while in multicellular organisms it is the basis of growth and repair.

Section 9.1   The Cell Cycle, *pp. 148–150*

Essential Knowledge Covered:
3.B.2

Vocabulary

apoptosis
cell cycle
chromatid
cyclins
cytokinesis
interphase
growth factors
mitosis
mitotic spindle
signals
somatic

REVIEW IT

Using the vocabulary on the left, **fill in the blanks** below.

The _____ is made up of two portions, interphase and the mitotic stage.

During _____ the cell grows and prepares for nuclear division. Nuclear

division and the division of the cytoplasm, or _____, occurs during

_____. In eukaryotic cells, the nucleus contains chromosomes. Each

chromosome has one DNA double helix or _____. Chromosomes are

distributed by the _____ during mitosis into two daughter cells. Agents which

influence the activities of a cell are called _____, such as _____

and _____. One signaling protein, p53, can stop the cycle if DNA is damaged

and bring about cell death or _____. Apoptosis decreases the number of

_____ cells.

What is the end result of mitosis?

_____

_____

USE IT

EK 3.B.2
Student Edition p. 148

Give an example of how cells control the cell cycle.

_____

_____

EK 3.B.2
Student Edition p. 148

Syndactyly is the condition of a human born with webbed or conjoined fingers. What do you think is the cause of this defect?

_____

_____

Section 9.1   The Cell Cycle (continued)

SUMMARIZE IT

EK 3.B.2

Identify the major stages or checkpoints of the cell cycle.

Definition	Stage	Checkpoint
Cell growth stage, before DNA replication occurs.		
		G_1
Cells grow and DNA replicates		
	G_2	
		G_2
Mitotic stage – cell division and cytokinesis occurs.		
		M

Section 9.2   The Eukaryotic Chromosome, *pp. 151–152*

EXTENDING KNOWLEDGE

This section takes the AP Essential Knowledge you have learned further, and may provide illustrative examples useful for the AP Exam.

REVIEW IT

Vocabulary

euchromatin
heterochromatin

Compare and contrast euchromatin and heterochromatin.

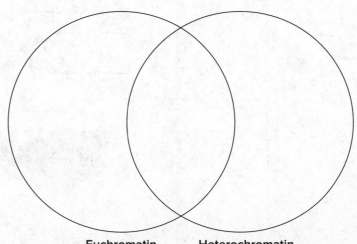

Euchromatin   Heterochromatin

USE IT

Student Edition p. 151

What are histones?

_____

_____

Student Edition p. 152

Describe one advantage that chromosomes may have in condensing before cell division.

_____

_____

SUMMARIZE IT

In the diagram below, **label** the chromosome, the histones, and the euchromatin.

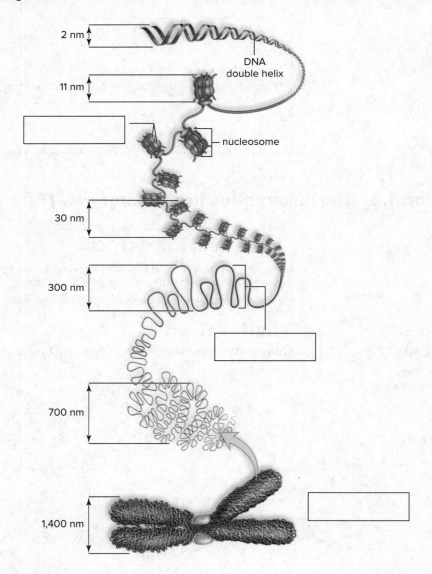

Section 9.3   Mitosis and Cytokinesis, *pp. 152–157*

Essential Knowledge Covered:
3.A.2

Vocabulary

cell plate
centriole
centromere
centrosome
chromatin
cleavage furrow
kinetochores

REVIEW IT

Given the definition on the left, **fill in** the correct structure on the right.

Definition	Structure
The tangled mass of DNA and proteins that make up a chromosome	
A newly formed plasma membrane that expands outward and fuses with an older membrane	
The microtubule-organizing center of the cell	
Protein complexes that develop on either side of the centromere during cell division	
Regions where sister chromatids are attached	
Barrel-shaped organelles in a centrosome of an animal	
An indentation of the membrane between two daughter cells in an animal	

USE IT

EK 3.A.2
Student Edition p. 152

The mosquito, *Aedes aegypti*, has a chromosome number of 6. **Identify** which mosquito cell is diploid and which is haploid.

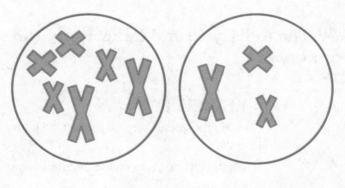

_____   _____

EK 3.A.2
Student Edition p. 152

What is an example of a haploid cell in animals?

_____

Section 9.3   Mitosis and Cytokinesis (continued)

EK 3.A.2
Student Edition p. 153

Draw a picture of a duplicated chromosome. **Label** the chromatids, the centromere, and the kinetochore.

SUMMARIZE IT

EK 3.A.2

Compare and contrast cytokinesis in animal and plant cells.

_____

_____

_____

_____

EK 3.A.2

What are stem cells and why are they important?

_____

_____

_____

Section 9.4   The Cell Cycle and Cancer, *pp. 158–160*

Essential Knowledge Covered:
3.A.2, 3.B.2, 3.D.3, 3.D.4

REVIEW IT

Using the vocabulary on the left, **fill in the blanks** below.

Vocabulary

angiogenesis
benign
cancer
malignant
metastasis
oncogenes
proto-oncogenes
telomeres
tumor
tumor suppressor genes

When cells begin to divide uncontrollably, an organism has _____. If cancer does not grow larger, it is considered _____, but if it has the ability to spread, it is _____. Abnormal cancer cells tend to pile up on each other to form a _____. Cancer cells may spread through the blood and lymph to start tumors elsewhere, a process called _____. Some tumor cells can direct the growth of new blood vessels in a process called _____. Two types of genes are often affected when cancer occurs: _____ and _____. Proto-oncogenes become _____ when they become mutated and cause cancer. Mutations in the ends of the chromosomes or _____ may also cause cancer.

Section 9.4 The Cell Cycle and Cancer (continued)

USE IT

EK 3.A.2
Student Edition p. 158

Determine if the following characteristic belongs to a cancer cell (CC) or a normal cell (NC).

Undergoes apoptosis ___

Undergoes metastasis ___

Has no contact inhibition ___

Differentiates ___

Has an abnormal nuclei ___

EK 3.D.4
Student Edition pp. 159–160

Suppose a person spends a lot of time applying a hazardous pesticide to a garden and then finds out that they have cancer. **Describe** a mutation to a proto-oncogene that might have been influenced by the presence of this pesticide to cause this disease.

_____

_____

_____

EK 3.B.2, 3.D.3, 3.D.4
Student Edition pp. 159–160

The person from the question above found that the cancer was actually caused by the failure of the tumor suppressor gene, p53. How is p53 related to cancer?

_____

_____

_____

SUMMARIZE IT

EK 3.B.2, 3.D.3, 3.D.4

Draw a diagram showing how a cancer cell may start on the skin, become a tumor, and invade the lymphatic and blood vessel system. Be sure to label your diagram.

Section 9.5   Prokaryotic Cell Division, *pp. 161–162*

Essential Knowledge Covered:
3.A.2

Vocabulary

asexual reproduction
binary fission
nucleoid

REVIEW IT

Fill in the chart below.

Process or Structure	Definition
Binary Fisson	
	Cellular divison in which offspring are genetically identical to parent
	A circular loop of DNA and protiens in a prokaroyte that is not enclosed by a membrane

EK 3.A.2
Student Edition pp. 161–162

Determine whether or not the following statements are true or false (T/F) about the prokaryotic cell division.

____ Prokaryotic offspring are genetically identical to the parent.

____ Prokaryotes lack a nucleus and other membranous organelles.

____ Prokaryotes do not have a chromosome.

____ After DNA replication occurs in prokaryotes, there are two genetically distinct chromosomes present.

Section 9.5   Prokaryotic Cell Division (continued)

EK 3.A.2
Student Edition p. 161

USE IT

Using the diagram below, **describe** the steps of prokaryotic division.

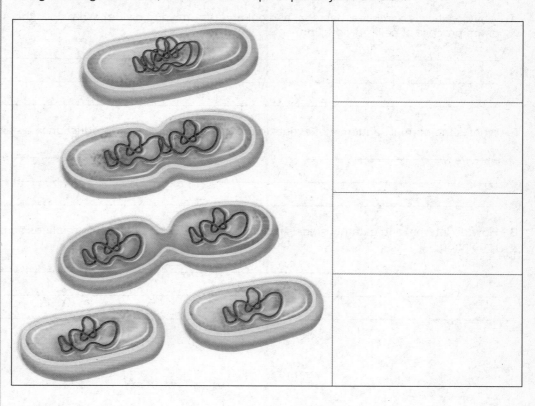

SUMMARIZE IT

EK 3.A.2

Compare and contrast eukaryotic and prokaryotic cell division.

_____

_____

_____

_____

Reviewing the Essential Questions

These questions were posed in the *Biology* chapter opener (page 147). Answer them using the knowledge you've gained from this chapter.

1. Why do all cells—archaea, bacteria, and eukaryotes—have to divide? What does this suggest about the evolution of the process of cell reproduction?

   _____

   _____

   _____

2. What is the normal sequence of events in the process of cellular reproduction in a eukaryotic cell?

   _____

   _____

   _____

3. How do internal and external signals regulate the cell cycle? What is the relationship between cancer and this regulation?

   _____

   _____

   _____

10 Meiosis and Sexual Reproduction

As you work through your AP Focus Review Guide, keep this chapter's Big Ideas in mind:

FOLLOWING THE BIG IDEAS

 BIG IDEA 1 The variation introduced during meiosis followed by fertilization plays an important role in evolutionary change.

 BIG IDEA 3 In sexually reproducing organisms, meiosis followed by fertilization recombines genetic information from both parents; changes in chromosome structure and number can have consequences for an individual's physiology.

 BIG IDEA 4 The variation produced by meiosis at the cellular levels affects all levels of an organism's physiology.

Section 10.1   Overview of Meiosis, *pp. 167–169*

Essential Knowledge Covered:
3.A.2, 3.C.2

Vocabulary

alleles
bivalent
diploid
haploid
homologues
gametes
meiosis
sexual reproduction
synapsis
synaptonemal complex
zygote

REVIEW IT

Given the definition on the left, **fill in** the correct vocabulary word on the right.

Definition	Vocabulary Word
Two homologous chromosomes in close association during meiosis I	
Reproductive cells	
A diploid cell	
A single set of chromosomes	
Homologous chromosomes come together and line up side by side	
The total number of chromosomes	
Reduces the chromosome number from diploid to haploid	
The process in which a synaptonemal complex forms	
The members of a pair of chromosomes	
Alternate forms of a gene	
Where haploid gametes merge into a zygote	

Section 10.1   Overview of Meiosis (continued)

USE IT

EK 3.A.2
Student Edition p. 167

Identify each chromosome, the homologous pair, the sister chromatids, and the nonsister chromatids.

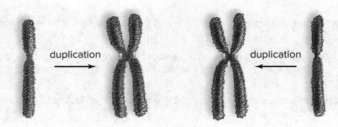

EK 3.A.2
Student Edition p. 167

Place an X under the cell where the following statement applies.

Statement	Zygote	Gamete
Has the haploid number of chromosomes		
Has a diploid number of chromosomes		
Undergoes development to become a mature organism		
Undergoes sexual reproduction to become a diploid cell		

EK 3.C.2
Student Edition p. 167

What are alleles? And where can they be found?

_____

_____

SUMMARIZE IT

EK 3.A.2

What is the central purpose of meiosis?

_____

_____

Section 10.1   Overview of Meiosis (continued)

EK 3.A.2

In the flow diagram below, **draw** the chromosomes of a cell (chromosome number = 6) undergoing DNA replication and producing daughter cells. **Label** when the cell is in meiosis I, and in meiosis II.

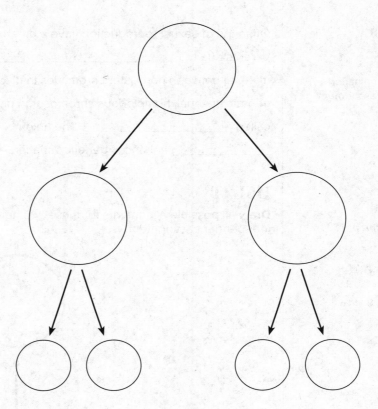

If the organism has a chromosome number of 6, how many daughter chromosomes will be present in the haploid cells?

_____

Section 10.2   Genetic Variation, *pp. 169–171*

Essential Knowledge Covered:
1.A.2, 3.A.2, 3.C.2

Vocabulary

crossing-over
fertilization
genetic recombination
independent assortment

EK 3.A.2
Student Edition p. 170

REVIEW IT

Using the vocabulary on the left, **fill in the blanks** to review the Learning Outcomes of this section.

Offspring of sexual reproduction have a different set of alleles and genes than their parents due to _____. _____ is the exchange of genetic material between non-sister chromatids in places called chiasmata. After exchanging genetic material, homologous chromosome pairs separate randomly during a process called _____. The union of male and female gametes, or _____, enhances genetic variation.

USE IT

Draw all possible alignments for these two homologous chromosomes during independent assortment.

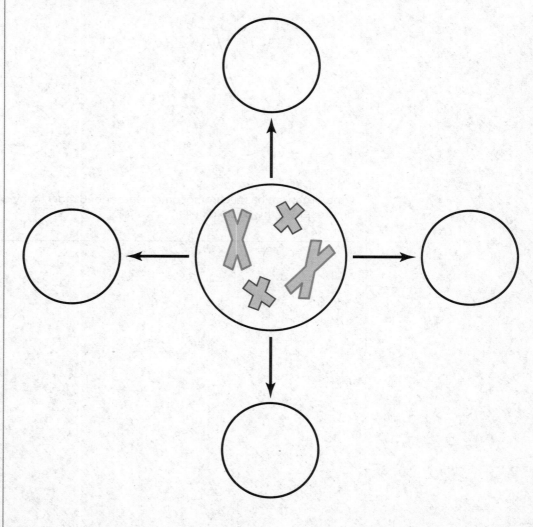

Section 10.2 Genetic Variation (continued)

SUMMARIZE IT

EK 1.A.2

How do organisms benefit from sexual reproduction in a changing environment?

_____

_____

_____

EK 3.C.2

Compare and contrast crossing-over and independent assortment.

_____

_____

_____

_____

Section 10.3 The Phases of Meiosis, *pp. 172—173*

Essential Knowledge Covered:
3.A.2

Vocabulary

interkinesis
meiosis I
meiosis II

REVIEW IT

Meiosis consists of two consecutive phases, meiosis I and meiosis II.

Between these two stages, there is a short period of rest.

What is this process called?

_____

_____

Compare and contrast meiosis I and meiosis II.

_____

_____

_____

_____

Section 10.3 The Phases of Meiosis (continued)

USE IT

Place a '2n' in the diploid cells and a 'n' in the haploid cells.

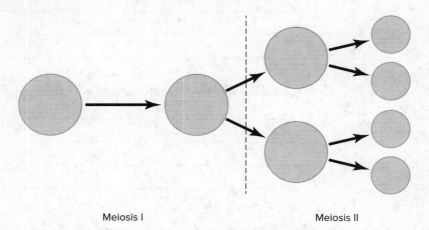

Meiosis I   Meiosis II

SUMMARIZE IT

EK 3.A.2

In your own words, **describe** the outcomes of the two phases of meiosis.

_____

_____

_____

Section 10.4 Meiosis Compared to Mitosis, *pp. 174–175*

Essential Knowledge Covered:
3.A.2

REVIEW IT

EK 3.A.2
Student Edition p. 174

List two similarities between mitosis and meiosis.

_____

_____

_____

EK 3.A.2
Student Edition p. 175

List the MAJOR differences between mitosis and meiosis.

_____

_____

_____

Section 10.4   Meiosis Compared to Mitosis (continued)

USE IT

EK 3.A.2
Student Edition pp. 174–175

Determine if the following statement applies to Mitosis or Meiosis.

Statement	Process
Daughter cells have half the chromosome number as the parent	
Daughter cells are genetically identical to each other and the parent	
After cytokinesis, there are two daughter cells	
Requires two nuclear divisions	
No pairing of the chromosomes occur	

Place an X under the phase where the following statement applies.

EK 3.A.2
Student Edition pp. 174–175

Statement	Meiosis I	Meiosis II	Mitosis
In the Prophase stage, there is a pairing of chromosomes			
In Anaphase, sister chromatids separate to become daughter cells			
Four haploid cells that are not genetically identical are formed			
Two diploid daughter cells that are identical to the parent are formed			

SUMMARIZE IT

EK 3.A.2
Student Edition p. 174

Why does mitosis produce daughter cells with the same chromosome number but meiosis results in a reduction?

_____

_____

Section 10.5   The Cycle of Life, *pp. 176–177*

Essential Knowledge Covered:
3.A.2

Vocabulary

gametogenesis
gametophytes
life cycle
polar body
oocytes
oogenesis
spermatogenesis
sporophytes

EK 3.A.2
Student Edition p. 176

REVIEW IT

Using the vocabulary on the left, **fill in the blanks** to review the Learning Outcomes of this section.

All reproductive events that occur from one generation to the next is called an organism's

_____. In humans, the only haploid phase of the life cycle occurs in the

gametes, whereas in plants, the haploid and diploid phases are broken into generations

known as _____ and _____. In humans, the production of gametes

or _____ is when meiosis occurs. Meiosis in males is known as

_____ and _____ in females. Oogenesis occurs in the ovaries,

more specifically in cells known as _____. A primary oocyte undergoes

meiosis I to produce a secondary oocyte and a _____.

What is the fate of the polar body?

_____

USE IT

Determine whether meiosis or mitosis occurs in the following human structure.

The Life Cycle of Humans		
Components	**Meiosis or Mitosis?**	**Specific Process**
Sperm		
Egg		
Zygote		—

Section 10.5   The Cycle of Life (continued)

EK 3.A.2
Student Edition p. 176

Insert the correct number in the statements about human spermatogenesis and oogenesis below.

Spermogenesis produces _____ viable sperm.

Oogenesis produces _____ egg and at least _____ polar bodies.

Each gamete has _____ chromosomes.

A zygote has _____ chromosomes.

SUMMARIZE IT

EK 3.A.2

Compare and contrast oogenesis and spermogenesis.

_____

_____

_____

Section 10.6   Changes in Chromosome Number and Structure, *pp. 177–182*

Essential Knowledge Covered:
3.A.3, 3.B.2, 3.C.1

REVIEW IT

Fill out the missing the vocab word or definition on the chart below.

Vocabulary

aneuploidy
Barr body
euploidy
karyotype
nondisjunction

Vocabulary Word	Definition
	A change in the number of chromosome number resulting from nondisjuction
karyotype	
	An inactived X chromosone
	The correct number of chromosones in a species
nondisjunction	

EK 3.A.3, 3.C.1
Student Edition p. 177

Name two aneuploid states and explain why the state occurs.

_____

_____

_____

USE IT

Given this chart of sex chromosome numbers, **identify** the syndrome.

Syndrome	Sex Chromosomes
	XXY
	XO
	XXX
	XYY

SUMMARIZE IT

EK 3.B.2

Describe four changes that can occur in chromosome structure which may lead to developmental abnormalities.

_____

_____

_____

AP Reviewing the Essential Questions

These questions were posed in the *Biology* chapter opener (page 166). Answer them using the knowledge you've gained from this chapter.

1. What are the similarities between meiosis and mitosis?

_____

_____

_____

_____

2. How does the process of meiosis reduce the chromosomes number from diploid to haploid?

_____

_____

_____

3. How does meiosis followed by fertilization increase genetic diversity?

_____

_____

_____

11 Mendelian Patterns of Inheritance

As you work through your AP Focus Review Guide, keep this chapter's Big Ideas in mind:

FOLLOWING THE BIG IDEAS

BIG IDEA 1 Inheritance of genes within a population is a cornerstone of species' ability to change over time.

BIG IDEA 3 Gregor Mendel's scientific approach allowed him to establish the basic principles of heredity.

Section 11.1   Gregor Mendel, *p. 187*

Essential Knowledge Covered:
3.A.3

REVIEW IT

EK 3.A.3
Student Edition p. 187

Name two reasons why Mendel was a successful scientist.

_____

_____

_____

USE IT

EK 3.A.3
Student Edition p. 187

Name two laws of genetics that Mendel proposed.

_____

_____

_____

SUMMARIZE IT

EK 3.A.3

In what way did Mendel's particulate theory disprove the blending concept?

_____

_____

_____

Section 11.2 Mendel's Laws, *pp. 188—194*

Essential Knowledge Covered:
3.A.3

EK 3.A.3
Student Edition pp. 188—192

REVIEW IT

Match the vocabulary terms with their proper definitions.

monohybrid cross	set of alleles an organism carries
allele	each individual has two factors for each trait and these factors separate during the formation of gametes
genotype	two different alleles
phenotype	cross of a single trait with organisms that are hybrid
Punnett square	an organism's physical appearance
law of segregation	a chart for determining the results of a test cross.
homozygous	two identical alleles
heterozygous	alternative versions of a gene
law of independent assortment	cross of two traits with organisms that are hybrid for both
dihybrid cross	the location of a gene on a chromosome
locus	each pair of factors segregates independently of the other pair, and all possible combinations of factors can occur in gametes

USE IT

EK 3.A.3
Student Edition p. 193

Identify if the pea plant is homozygous or heterozygous for being tall (T).

TT _____

Tt _____

tt _____

Section 11.2 Mendel's Laws (continued)

EK 3.A.3
Student Edition p. 193

Two parents both carry a recessive and dominant allele for the recessive genetic disorder cystic fibrosis (Cc). If they had a child with cystic fibrosis, what genotype would this child have? What is the probability they will have a child with cystic fibrosis? Use a Punnett square to **illustrate** your answer.

SUMMARIZE

EK 3.A.3

In a particular strain of roses, plants with pink flowers (P) are dominant over those with white flowers (p). If you wanted to determine the genotype of a rose plant that produced pink flowers, what test would you run?

_____

If the results of the test gave you 50% white roses and 50% roses, what was the genotype of the original rose?

_____

Suppose the white rose plants also carry a recessive gene for having no thorns (t). If a white thorn-less rose plant (wwtt) was crossed with a heterozygous pink rose plant (WwTt), what would the percentage of pink thorn-less roses be? Use a Punnett square to help you **determine** your answer.

What is the phenotypic ratio of this cross? _____

Section 11.3 Mendelian Patterns of Inheritance and Human Disease, *pp. 194–197*

Essential Knowledge Covered:
3.A.3

REVIEW IT

EK 3.A.3
Student Edition pp. 194–195

Fill in the correct word to complete the following sentences.

An _____ is any chromosome other than a sex chromosome.

A _____ is a person who appears normal but is able to have a child with a genetic disorder.

USE IT

EK 3.A.3
Student Edition p. 195

Compare and contrast what it means for a disorder to be autosomal dominant versus autosomal recessive.

_____

_____

_____

_____

_____

_____

EK 3.A.3
Student Edition pp. 196–197

Review if the disorder you learned about in this section is either autosomal recessive (R) or autosomal dominant (D).

Disorder	
cystic fibrosis	____
phenylketonuria	____
osteogenesis imperfecta	____
Huntington disease	____
methemoglobinemia	____
hereditary spherocytosus	____

SUMMARIZE IT

EK 3.A.3

Examine this pedigree below. Do the affected individuals carry an autosomal dominant or recessive disorder?

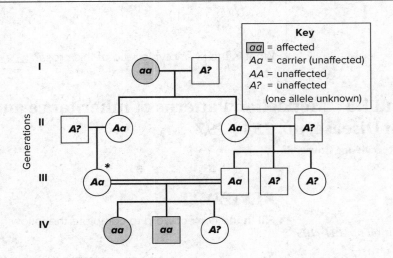

Section 11.3   Mendelian Patterns of Inheritance and Human Disease (continued)

EK 3.A.3

The double line in generation III denotes what? How does this influence the probability of inheriting a harmful allele?

_____

_____

_____

If this disorder charted in the pedigree above was methemoglobinemia, what could you tell about the blood of the affected individuals?

_____

_____

_____

How is it possible that someone that might appear normal may still produce a child with the trait?

_____

Section 11.4   Beyond Mendelian Inheritance, *pp. 197–204*

Essential Knowledge Covered:
3.A.2, 3.A.3

REVIEW IT

Using the vocabulary, **identify** which complex pattern of inheritance explains the traits in the organism below.

Vocabulary

codominance
incomplete dominance
incomplete penetrance
multifactorial traits
pleiotropy
polygenic inheritance
X-linked

Organism	Trait	Pattern of Inheritance
Fruit fly	Only males have white colored eyes	
Human	Marfan syndrome effecting lungs, eyes, and blood vessels	
Human	Has both A and B antigen on red blood cells	
Human	Skin color in the middle range	
Four-o-clock plants	Red and white flowered parents produce 1:2:1 red: pink:white flowered offspring	
Human	Polydactyly inherited in an autosomal dominant manner from a parent with only 10 digits	
Himalayan rabbit	Has dark fur only at the extremities	

Section 11.4   Beyond Mendelian Inheritance (continued)

EK 3.A.3
Student Edition p. 202

USE IT

A man is considered hemizogous for his red-green color blindness.
What does this mean?

_____

_____

EK 3.A.2

SUMMARIZE IT

A person is suffering from sickle-cell anemia. Describe the disadvantages and
advantages of the mutation they carry.

_____

_____

_____

_____

AP Reviewing the Essential Questions

These questions were posed in the *Biology* chapter opener (page 186). Answer them using the knowledge
you've gained from this chapter.

1. What is the relationship between genes and their passage from parent to offspring to natural selection and
   evolution?

   _____

   _____

   _____

2. How does the behavior of chromosomes during meiosis explain Mendel's laws of segregation and independent
   assortment?

   _____

   _____

   _____

3. How does an understanding of Mendelian genetics help us understand the link between genes and human
   genetic diseases?

   _____

   _____

   _____

12 Molecular Biology of the Gene

As you work through your AP Focus Review Guide, keep this chapter's Big Ideas in mind:

FOLLOWING THE BIG IDEAS

 BIG IDEA 1 DNA stores and transmits genetic information in all organisms.

 BIG IDEA 3 Genetic information in the form of a coded sequence of nucleotides dictates the sequence of amino acids which will make a protein.

Section 12.1 The Genetic Material, *pp. 208–212*

Essential Knowledge Covered:
1.B.1, 3.A.1

EK 3.A.1
Student Edition pp. 208–210

REVIEW IT

Identify the major components of DNA.

Definition	Component(s)
The two purines bases	
The two pyrimdine bases	
The shape of DNA	
The molecules that make up the "rungs" of the double helix	

Name three scientists who helped determine the structure of DNA.

_____

USE IT

EK 1.B.1
Student Edition p. 210

List Chargaff's two rules.

1. _____

2. _____

Section 12.1   The Genetic Material (continued)

EK 3.A.1
Student Edition pp. 208–209

On the timeline below, **describe** the important studies these scientists did to aid in identifying DNA as heritable material.

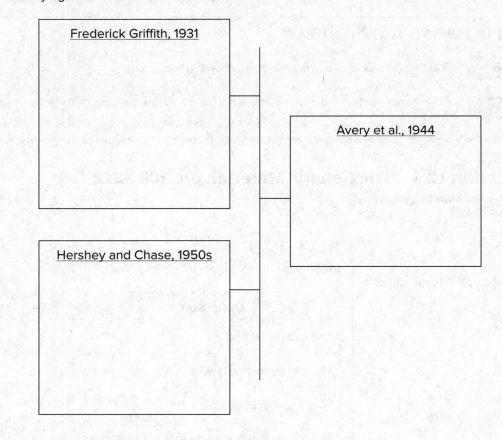

Frederick Griffith, 1931

Avery et al., 1944

Hershey and Chase, 1950s

EK 3.A.1
Student Edition p. 212

Describe the structure of DNA.

_____

_____

SUMMARIZE IT

EK 1.B.1

Explain how it is possible that, although only one of four bases is possible at each nucleotide position in DNA in all species, there is so much variability in genetic material.

_____

_____

_____

Section 12.2 Replication of DNA, *pp. 213–215*

Essential Knowledge Covered:
3.A.1, 3.C.1

Vocabulary

DNA helicase
DNA ligase
DNA primase
semiconservative replication
single-stranded binding
  proteins
template

REVIEW IT

Use the vocabulary words on the left to review the process of DNA replication.

The process of DNA replication is called _____ because the daughter strand contains one old strand and one new strand. The parental double helix serves as a _____ for the new strand. DNA replication begins with DNA being unwound by the enzyme _____. _____ attach to the open strands to prevent DNA from recoiling so that _____ can place a short primer on the strand that needs to be replicated. In eukaryotes, wherever DNA replication begins is called a _____. _____ binds the fragments of the replicated lagging strands back to together.

Help fill out the missing the enzyme or function on the chart below.

DNA Replication	
Structure	**Function**
	Places primers, begins DNA synthesis, and proof reads strands
DNA helicase	
	Mends Okazaki fragments together

USE IT

EK 3.A.1
Student Edition p. 214

What does it mean that DNA synthesized in the 5' to 3' direction?

_____

_____

EK 3.A.1
Student Edition p. 214

Draw a diagram showing the enzymes involved in eukaryotic DNA replication in action. Be sure to include the following structures in your answer.

Structure
DNA polymerase
5' strand
3' strand
Okazaki fragments
DNA helicase
SSBs
DNA primase
A primer
DNA ligase

Section 12.2   Replication of DNA (continued)

SUMMARIZE IT

EK 3.A.1

Compare and contrast prokaryotic and eukaryotic DNA replication.

_____

_____

_____

_____

EK 3.C.1

How does DNA polymerase aid in accurately copying DNA? What would happen if it couldn't?

_____

_____

_____

Section 12.3   The Genetic Code of Life, *pp. 215–217*

Essential Knowledge Covered:
1.B.1, 3.A.1

REVIEW IT

In this section, you were introduced to three different types of RNA: mRNA, tRNA, and rRNA. What makes them similar? What makes them different? **Fill out** the chart below.

	Similar Traits	Unique Function
mRNA		
tRNA		
rRNA		

EK 1.B.1
Student Edition p. 216

Illustrate how DNA flows from the nucleus to become a polypeptide. Be sure to include the words *transcription*, *translation*, and *mRNA* in your answer.

USE IT

EK 3.A.1
Student Edition p. 216

What is the central dogma?

_____

_____

Section 12.3   The Genetic Code of Life (continued)

List the three stop codons and the one start codon.

_____

_____

SUMMARIZE IT

EK 3.A.1

Describe the two steps of how DNA is converted into a proteins.

_____

_____

EK 1.B.1

How does the universal nature of the genetic code link all living organisms?

_____

_____

Section 12.4   First Step: Transcription, *pp. 217–220*

Essential Knowledge Covered:
3.A.1, 3.B.1

REVIEW IT

Given the definition on the left, **fill in** the correct structure on the right.

Vocabulary

exons
introns
mRNA transcript
promoter
ribozyme
RNA polymerase

Definition	Structure
A molecule released by RNA polymerase	
Non-protein coding regions of pre-mRNA	
Attaches to the promoter and joins nucleotides together	
Protein-coding regions of pre mRNA	
A region of DNA that defines the start of transcription	
An enzyme made of RNA rather than protein	

USE IT

EK 3.A.1
Student Edition p. 218

Identify the three major modifications of mRNA that occur before it is ready to leave the nucleus for protein translation.

_____

_____

Section 12.4   First Step: Transcription (continued)

EK 3.A.1
Student Edition p. 219

Describe two key function of introns.

_____

_____

_____

EK 3.B.1
Student Edition p. 217

Name three reasons why the promoter important to DNA transcription.

_____

_____

SUMMARIZE IT

EK 3.A.1

Process this pre-mRNA so it can go deliver a message!

EXON	INTRON	EXON	INTRON

Section 12.5   Second Step: Translation, *pp. 220–225*

Essential Knowledge Covered:
3.A.1

REVIEW IT

EK 3.A.1
Student Edition pp. 222–224

Identify the three major steps of protein synthesis.

Definition	Phase
The step that brings all the translation components together	
Polypeptides increase in length, one amino acid at a time	
Finished polypeptide and assembling components separate	

Determine if the following statements refer to a transfer RNA (tRNA) or ribosomal RNA (rRNA).

_____ Transfers amino acids to the ribosomes

_____ Is packaged in two subunits of unequal size

_____ Becomes "charged" when an amino acid is attached

_____ Contains an anticodon

_____ Aids in the creation of a peptide bond between animo acids

USE IT

EK 3.A.1
Student Edition pp. 220–221

How does a tRNA molecule correctly align amino acids in the order predetermined by DNA?

_____

_____

_____

EK 3.A.1
Student Edition p. 221

What are the three binding sites on the small subunit of the ribosome and what role do they play in polypeptide synthesis?

_____

_____

_____

SUMMARIZE IT

EK 3.A.1

How does termination occur?

_____

_____

EK 3.A.1

Illustrate what happens to mRNA once it moves into the cytoplasm and how it becomes a protien. Be sure to include tRNA, ribosomes, and anticodons in your drawing.

AP Reviewing the Essential Questions

These questions were posed in the *Biology* chapter opener (page 207). Answer them using the knowledge you've gained from this chapter.

1. How did historical experiments help identify DNA as the carrier of genetic information?

   _____

   _____

   _____

2. How does the chemical structure of DNA as defined by the Watson and Crick model determine DNA's ability to store and transmit genetic information?

   _____

   _____

   _____

3. How does genetic information stored in DNA flow from a sequence of nucleotides in a gene to a sequence of amino acids in a protein?

   _____

   _____

   _____

13 Regulation of Gene Expression

As you work through your AP Focus Review Guide, keep this chapter's Big Ideas in mind:

FOLLOWING THE BIG IDEAS

 BIG IDEA 1 Random changes in DNA can lead to phenotypic variation that can be acted upon by natural selection.

 BIG IDEA 3 Regulation of prokaryotic genes is often at the operon level, while eukaryotes fine tune gene expression and modify gene products.

BIG IDEA 4 Regulation of gene activity involves intricate interactions with internal and external factors.

Section 13.1 Prokaryotic Regulation, *pp. 229–231*

Essential Knowledge Covered:
3.B.1, 3.B.2, 4.A.2

Vocabulary

compressor
inducer
operator
promoter
regulator
structural genes

REVIEW IT

Given the definition on the left, **fill in** the correct vocabulary word on the right.

Definition	Vocabulary Word
An enzyme that brings out the expression of a gene	
A short portion of DNA located before the structural genes	
An area outside the operon which encodes for a repressor protein	
Genes that code for enzymes and proteins involved in a the metabolic pathway of an operon	
A molecule that binds to a repressor and prevents a gene from being expressed	
A short sequence of DNA where RNA polymerase first attaches to begin the transcription of an operon	

EK 3.B.1, 3.B.2
Student edition pp. 229–230

List the four parts of an operon.

_____

_____

Section 13.1 Prokaryotic Regulation (continued)

EK 3.B.1, 3.B.2, 4.A.2
Student Edition pp. 229–231

USE IT

The following diagram illustrates the *trp* operon which regulates gene expression in *E. coli* with the amino acid tryptophan. Study the diagram below then answer the following questions.

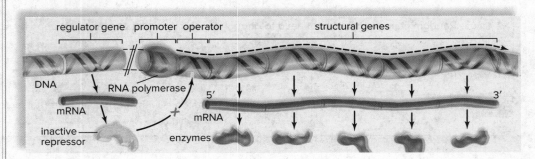

What are the products that this operon regulates?

_____

Is this operon in the "on" or "off" position (i.e. allowing for gene expression or shutting off the expression of gene expression) without tryptophan present.

_____

Where does tryptophan bind?

_____

What happens to the operator and promoter in the operon when tryptophan binds?

_____

_____

Why is the *trp* operon considered a repressible operon?

_____

_____

SUMMARIZE IT

EK 3.B.2

Compare and contrast the trp operon and the lac operon.

_____

_____

_____

_____

_____

Section 13.2   Eukaryotic Regulation, *pp. 232–238*

Essential Knowledge Covered:
3.B.1, 3.B.2, 3.C.1, 4.A.2

REVIEW IT

EK 3.B.2
Student Edition p. 232

List five mechanisms that can control gene expression in eukaryotes.

_____

_____

USE IT

EK 3.B.1
Student Edition pp. 232–238

Based on the description of genetic control, **determine** the name of the control and where it takes place.

Name of control	Description	Location it takes place
	A missing transcription factor	
	Incorrect folding	
	A missing 5' cap	
	pre-mRNA is spliced alternatively	

EK 3.B.1, 4.A.2
Student Edition pp. 232–238

How does a protein enter a proteasome and what happens once it enters?

_____

_____

_____

SUMMARIZE IT

EK 3.C.1

How might scientists be able to use sRNA as therapeutic agents to stop certain diseases?

_____

_____

_____

Essential Knowledge Covered:
1.C.3, 3.B.1, 3.C.1

EK 1.C.3, 3.C.1
Student Edition p. 238

EK 3.B.1, 3.C.1
Student Edition pp. 239–240

EK 3.C.1
Student Edition pp. 238–239

EK 3.B.1

REVIEW IT

Define gene mutation.

_____

List two causes of mutation.

_____

USE IT

How might a base substitution in a hemoglobin gene cause a red blood cell to become sickle shaped?

_____

_____

Place an X under the mutation where the following statement may apply.

	Induced	**Spontaneous**	**Point**	**Frameshift**
Happens for no apparent reason				
Occurs when a nucleotide is added or deleted from DNA				
Results from exposure to chemicals or radiation				
Is a permeant change in the sequence of DNA				
Involve a change of a single nucleotide				

SUMMARIZE IT

Compare and contrast how mutations in a tumor suppressing gene and a proto-oncogene can lead to cancer.

_____

_____

_____

_____

Reviewing the Essential Questions

These questions were posed in the *Biology* chapter opener (page 228). Answer them using the knowledge you've gained from this chapter.

1. How do changes in DNA lead to variation in phenotypes that are subject to natural selection?

   _____

   _____

   _____

2. How does the operon model explain how gene expression is regulated in prokaryotes?

   _____

   _____

   _____

3. How do regulatory genes, molecules, and transcription factors control gene expression in eukaryotes?

   _____

   _____

   _____

4. How can cells specialize when they contain the same set of genetic instructions?

   _____

   _____

   _____

14 Biotechnology and Genomics

As you work through your AP Focus Review Guide, keep this chapter's Big Ideas in mind:

FOLLOWING THE BIG IDEAS

 The field of comparative genomics is yielding valuable new insights into the relationships between species, impacting taxonomy and evolutionary biology.

 Genetic engineering allows beneficial changes to be made in DNA and RNA sequences, improving the products or the actions of these molecules.

Section 14.1   DNA Cloning, *pp. 245–247*

Essential Knowledge Covered:
3.A.1

REVIEW IT

Given the definition on the left, **identify** the correct tool on the right.

Vocabulary

cloning
DNA fingerprinting
gel electrophoresis
gene therapy
polymerase chain reaction
recombinant DNA

Technology	Vocabulary Word
Can identify and distinguish between individuals bases on variation in their DNA	
Using cloned genes to modify a person	
Amplifies a small piece of DNA by making millions of copies using DNA polymerase	
Producing genetically identical copies of DNA, cells, or organisms through asexual means	
Contains DNA from two or more different sources and is carried in a vector	
Separates DNA fragments accordin	

SUMMARIZE IT

Someone stole the cookies from the cookie jar, but they made a critical mistake: they left their DNA behind all over an empty glass of milk at the scene of the crime. Taking cookie theft very seriously, the owner had a DNA profile run and tested against the DNA on the three suspects: Who Me, Yes You, and Couldn't Be. With this information in mind, answer the following questions.

EK 3.A.1
Student Edition p. 246

There was only a trace amount of DNA recovered from the empty glass. How did scientists increase the amount so they could perform more accurate tests on it? **Describe** the process in detail.

_____

_____

_____

_____

The scientist working on the case only amplified a portion of the DNA which codes for the human CookieMnstr gene. What tool did the scientist use to do this?

_____

Draw a line under all the biotechnology tools used in the next sentence:
The CookieMnstr gene was broken into short tandem repeat sequences using a restriction enzyme and separated using gel electrophoresis for analysis.

Looking at the results, who did it and how can you tell?

Base repeat units	Glass DNA	Who Me	Yes You	Couldn't Be
	▓		▓	▓
		▓		
	▓	▓	▓	▓
	▓	▓	▓	
				▓

DNA band pattern

Section 14.2 Biotechnology Products, *pp. 247–249*

Essential Knowledge Covered:
1.C.3, 3.A.1

Vocabulary

biotechnology products
gene pharming
GMOs

EK 1.C.3, 3.A.1
Student Edition p. 248

REVIEW IT

Define the following biotechnological terms.

Genetically Modified Organisms— _____

Biotechnology Products— _____

Gene Pharming— _____

USE IT

Determine if the following statements about biotechnology products are true (T) or false (F).

Bioengineering can enhance the natural abilities of bacteria to breakdown toxins for their use in bioremediation.	
Humans contain plasmid vectors which are directly inserted into bacteria for the production of human pharmaceuticals.	
Only animals can be cloned.	
Golden Rice was engineered to contain human genes.	
Animal cloning is a difficult process with a low success rate.	

Section 14.2   Biotechnology Products (continued)

EK 3.A.1

SUMMARIZE IT

Describe how scientists used mice to discover the function of the human section of DNA called *SRY*.

_____

_____

Section 14.3   Gene Therapy, *p. 250*

Essential Knowledge Covered:
3.A.1

EK 3.A.1
Student Edition p. 250

SUMMARIZE IT

Fill in the charts to describe the two forms of gene therapy.

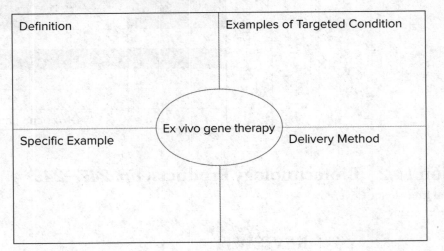

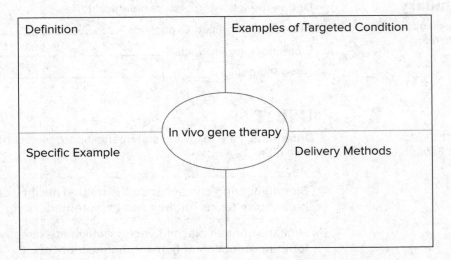

Section 14.4   Genomics, *pp. 251–255*

Essential Knowledge Covered:
3.A.1, 3.C.3

REVIEW IT

Using the vocabulary on the left, **match** the definition and the term.

Vocabulary

bioinformatics
comparative genomics
DNA microarrays
functional genomics
genetic profile
genomics
Human Genome Project
intergenic sequences
proteome
proteomics
structural genomics
tandem repeat
transposons

Term	Definition
bioinformatics	understanding the exact role of the genome in cells or organisms
comparative genomics	the application of computer technologies to the study of biological information
functional genomics	the study of the structure, function, and interaction of cellular proteins
structural genomics	repeated sequences next to each other on a chromosome
genomics	comparing the human genome to other organisms
proteome	the study of genomes
proteomics	the entire collection of a species' proteins
transposons	DNA sequences with the ability to move within and between chromosomes
tandem repeat	the order of the base pairs in the human genome
Human Genome Project	all the mutations in the genome of an individual
DNA mircoarrays	DNA sequences that occur between genes
genetic profile	knowing the sequence of the bases and how many genes an organism has
intergenic sequences	known DNA sequences in known locations on a slide or chip

Section 14.4   Genomics (continued)

EK 3.A.1
Student Edition pp. 251–255

USE IT

Identify the field you would need to solve the following question.

Question	Technology
What type of proteins are present in the human eye?	
How many genes are in the genome of a panda?	
What has more genes, an orchid or a potato?	
What genes in the brain of a mouse are turned on when it is sleeping?	

SUMMARIZE IT

EK 3.C.3
Student Edition p. 253

Describe how a transposon may turn the kernel of corn a different color.

_____

_____

_____

EK 3.A.1
Student Edition p. 251

It was once thought that the Human Genome Project would discover over 100,000 genes when the project first began but this estimate was revised to 21,000–23,000 as more data became available. What accounts for all the "missing genes"?

_____

_____

_____

AP Reviewing the Essential Questions

These questions were posed in the *Biology* chapter opener (page 244). Answer them using the knowledge you've gained from this chapter.

1. How can DNA analysis and genome comparison allow us to better understand the evolution of species?

_____

_____

_____

2. How can genetic engineering techniques be used to benefit human society?

_____

_____

_____

3. How can the manipulation of DNA by humans affect the evolution of species, and what are the ethical, medical, or social implications?

_____

_____

_____

15 Darwin and Evolution

As you work through your AP Focus Study Guide, keep this chapter's Big Ideas in mind:

FOLLOWING THE BIG IDEAS

 Darwin's theory of natural selection states that organisms survive because they possess more favorable adaptations to their environment, and these heritable traits are passed on to offspring.

 Evolution by natural selection comes about from interaction between an organism and its environment.

Section 15.1   History of Evolutionary Thought, *pp. 262–264*

Essential Knowledge Covered:
1.A.1

Vocabulary

catastrophism
evolution
extant
inheritance of acquired
    characteristics
paleontology
strata
uniformitarianism
vestigial structures

REVIEW IT

Given the definition on the left, **fill in** the vocabulary word concerning the history of evolutionary thought on the right.

Definition	Vocabulary Word
Cuvier's idea that worldwide catastrophes had occurred and God had repopulated the world with new species	
Still in existence	
Anatomical structures that functioned in an ancestor but have since lost most or all of their function in a descendant	
The study of fossils	
Darwin's idea that genetic change occurs in species over time due to natural forces	
Different layers of sediment	
Lyell's idea that natural processes today are the same as those which occurred in the past	
Lamarck's idea that the environment can induce physical changes in an animal over its lifetime that become heritable	

EK 1.A.1
Student Edition pp. 263–264

USE IT

Describe why paleontology and geology are important to understanding evolution.

_____

_____

Section 15.1 History of Evolutionary Thought (continued)

EK 1.A.1
Student Edition pp. 263—264

How did Lamarck's theory influence Darwin's idea of natural selection?

_____

_____

_____

SUMMARIZE IT

EK 1.A.1

Why are grass snakes green? Give an example of how Cuvier, Lamarck, and Darwin might answer this question.

_____

_____

_____

EK 1.A.1

How did Darwin use Thomas Malthus' work to describe changes in animal population?

_____

_____

_____

Section 15.2 Darwin's Theory of Evolution, *pp. 265—270*

Essential Knowledge Covered:
1.A.1, 1.A.2, 1.A.4, 1.C.3, 4.B.4

REVIEW IT

Identify the vocabulary word which the example illustrates.

Vocabulary

adaptation
artificial selection
biogeography
fitness

Vocabulary Word	Example
	A toad has the ability to look like a stone and blend in with the forest floor
	Darwin studied animals in the Southern Hemisphere and in the Northern Hemisphere
	Gray mice living on gray rocks have the greatest number of offspring which survive
	A golden retriever has big floppy ears and a waggy tail

EK 1.A.1, 1.A.4
Student Edition pp. 265—266

List the four observations that natural selection is based upon.

_____

_____

_____

Section 15.2   Darwin's Theory of Evolution (continued)

USE IT

EK 1.A.2
Student Edition p. 270

A doctor prescribes you an antibiotic for a bacterial infection and says, "It's very important that you finish the entire course, even if you think you feel better." What is the evolutionary concern?

_____

_____

EK 1.A.2
Student Edition p. 270

The Industrial Revolution in Great Britain caused a lot of smog and soot to pollute the air. How did this effect the populations of peppered moths living in the area?

_____

_____

EK 1.A.4, 4.B.4
Student Edition p. 265

Describe two reasons why studying geology helped Darwin form his theory of evolution.

_____

_____

EK 1.A.1, 1.A.2
Student Edition pp. 267–268

Compare and contrast artificial and natural selection.

_____

_____

_____

SUMMARIZE IT

EK 1.A.1, 1.C.3

Darwin studied many organisms on his trip aboard the HMS Beagle, including groups of finches. The finches all had very different beaks. **Use** the graph below to describe one possible way in which beaks may change in a finch.

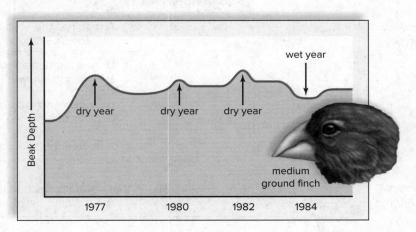

_____

_____

_____

Section 15.3   Evidence for Evolution, *pp. 270–275*

Essential Knowledge Covered:
1.A.1, 1.A.4

Vocabulary

analogous
fossils
homeobox
homologous
transitional fossils

REVIEW IT

Give two examples of homologous structures.

_____

_____

Give two examples of analogous structures.

_____

_____

Name the type of genes which orchestra the development of the body plan in all animals. Are these genes homologous or analogous?

_____

Describe what a fossil is and what makes a fossil "transitional."

_____

_____

USE IT

EK 1.A.4
Student Edition pp. 270–271

Draw a picture showing how *Tiktaalik roseae* has transitional features between a fish and a tetrapod.

EK 1.A.4
Student Edition p. 274

Determine whether or not the following structures are homologous (H) or analogous (A).

____ The tail of an airplane and the tail of a bird

____ The tentacles of a star-nosed mole and the tentacles of an octopus

____ The bill of a duck and the beak of a sparrow

____ The bones in the fin of a whale and the bones in the leg of a cow

SUMMARIZE IT

EK 1.A.1

Describe how evolution can be observed and tested.

_____

_____

_____

Section 15.3 Evidence for Evolution (continued)

EK 1.A.1

A gardener we met back in Section 9.4 had been using pesticides. This particular garden had pesticides applied to it for many generations to control insects in their gardens. One day, the gardener noticed that the insects on the plants no longer died when the pesticide was applied. **Describe** what may have happened to the insect populations in terms of evolution.

_____

_____

_____

AP Reviewing the Essential Questions

These questions were posed in the *Biology* chapter opener (page 261). Answer them using the knowledge you've gained from this chapter.

1. How do random mutations in DNA and genetic variation result in phenotypic variations that are subject to natural selection?

_____

_____

_____

2. How does evolution by natural selection explain both the unity and diversity of life on Earth?

_____

_____

_____

3. How can environmental change and human activity impact the evolution of species?

_____

_____

_____

16 How Populations Evolve

As you work through your AP Focus Review Guide, keep this chapter's Big Ideas in mind:

FOLLOWING THE BIG IDEAS

 BIG IDEA 1   Microevolution, or evolution within populations, is measured as a change in allele frequencies over generations.

Section 16.1   Genes, Populations, and Evolution, *pp. 280–285*

Essential Knowledge Covered:
1.A.1, 1.A.3, 1.A.4

Vocabulary

allele frequency
assortative mating
bottleneck effect
founder effect
inbreeding
gene flow
gene pool
genetic drift
Hardy-Weinberg equilibrium
Hardy-Weinberg principle
nonrandom mating
population
population genetics
microevolution
reproductive isolation

REVIEW IT

Using the vocabulary on the left, **fill in the blanks** to review the Learning Outcomes of this section.

The field of biology that studies diversity of populations at the genetic level is known as _____. A _____ is a group of individuals of the same species living together in the same geographic region. Evolutionary change within populations is referred to as _____, and geneticists look for these changes in the alleles of all genes of individuals in a population or in the _____. The percentage of each allele in a population's gene pool is known as the _____. When this percentage is stable it is in _____. The frequency of a non-evolving population can be described by a mathematical model called the _____. The movement of alleles between populations is known as _____. If the movement of alleles of a population become more and more different over time, _____ can occur which means parts of the populations can no longer interbreed. A change in allele frequencies due to chance events is said to due to _____. Two types of genetic drift include the _____ and the _____. The mating between relatives or _____ does not affect the frequency of alleles alone but can have a significant impact on the genotype and phenotype of individuals. When individuals choose a mate with a preferred trait, this is known as _____ or, more specifically _____, and can skew the frequency of prevalent genotypes.

Section 16.1 Genes, Populations, and Evolution (continued)

EK 1.A.1, 1.A.4
Student Edition p. 282

USE IT

Describe the five conditions of the Hardy-Weinberg principle that need to be met in order for a population to be at equilibrium.

_____

_____

EK 1.A.1, 1.A.4
Student Edition p. 282

Estimate the genotype frequencies of a population at Hardy-Weinberg equilibrium given the following allele frequencies: 0.30 H, 0.70 h

If the above population's F_2 generation has a genotypic frequency of HH=0.7, Hh=0.22, hh=0.08, can you determine if evolution occurred?

_____

_____

SUMMARIZE IT

EK 1.A.3

Compare and contrast a bottleneck effect and a founder effect.

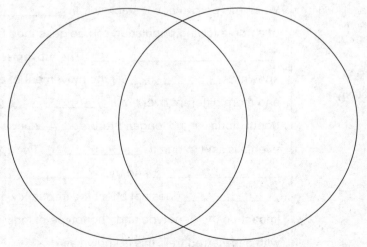

Bottleneck Effect   Founder Effect

Vocabulary

cost-benefit analysis
dominance hierarchies
fitness
polygenic
sexual dimorphism
sexual selection
territoriality
territory

EK 1.A.2
Student Edition p. 286

REVIEW IT

Fill out the missing the vocab word or definition on the chart below.

Vocabulary Word	Definition
cost-benefit analysis	
	the ability to produce surviving offspring
territory	
	adapative changes in males and females that lead to an increased abolity to secure a mate
sexual dimorphism	
dominance hierarchies	
	controlled by many genes
territoriality	

Describe the three types of natural selection shown below in terms of change in phentoypes.

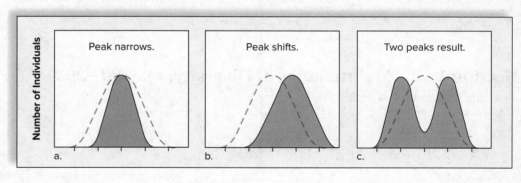

a. Peak narrows.   b. Peak shifts.   c. Two peaks result.

A. _____

B. _____

C. _____

Section 16.2   Natural Selection (continued)

USE IT

EK 1.A.1
Student Edition pp. 286–287

Identify the type of natural selection described in the examples below.

Human infants with an intermediate weight have a better chance of survival	
A population of fish feed on detritus in deep water but start feeding on flies when in shallow water	
Asters growing in a tundra have small leaves and flowers	
A drab female bird mates with a brightly colored male	

SUMMARIZE IT

EK 1.A.2

Explain in terms of natural selection, why British land snails tend to have two distinct phenotypes.

_____

_____

_____

EK 1.A.2

Describe two ways sexual selection may occur.

_____

_____

Section 16.3   Maintenance of Diversity, *pp. 290–293*

Essential Knowledge Covered:
1.A.1, 1.A.2

REVIEW IT

EK 1.A.1, 1.A.2
Student Edition p. 293

_____ occurs when the heterozygote is favored over the two homozygotes.

How does this affect the maintenance of diversity?

_____

_____

USE IT

EK 1.A.2
Student Edition p. 293

Describe another mechanism which promotes the maintenance of diversity.

_____

_____

_____

Section 16.3 Maintenance of Diversity (continued)

EK 1.A.1
Student Edition p. 292

SUMMARIZE IT

Explain why Sickle-cell disease remains prevalent in Africa, despite the devastating affects it can have on an individual.

_____

_____

_____

AP Reviewing the Essential Questions

These questions were posed in the *Biology* chapter opener (page 279). Answer them using the knowledge you've gained from this chapter.

1. What is the connection between change in the environment and change in allele frequencies?

_____

_____

_____

_____

2. How can the Hardy-Weinberg mathematical model be used to analyze genetic drift and effects of selection in the evolution of populations?

_____

_____

_____

_____

17 Speciation and Macroevolution

As you work through your AP Focus Review Guide, keep this chapter's Big Ideas in mind:

FOLLOWING THE BIG IDEAS

 BIG IDEA 1 Macroevolution, or the origin of new species, results from the accumulation of microevolutionary change over time.

Section 17.1 How New Species Evolve, *pp. 297–303*

Essential Knowledge Covered:
1.C.1, 1.C.3

REVIEW IT

Define the species concepts below.

Concept	Definition
Morphological Species Concept	
Evolutionary Species Concept	
Phylogenetic Species Concept	
Biological Species Concept	

List five prezygotic isolating mechanisms.

_____

_____

List two postzygotic mechanisms.

_____

_____

Section 17.1 How New Species Evolve (continued)

EK 1.C.1, 1.C.3
Student Edition pp. 297–298

USE IT

Draw a line to match the correct word with the correct definition.

Speciation	Evolution on a large scale
Cryptic species	The mating between two species
Macroevolution	Species which look almost identical but are very different in other traits
Diagnostic traits	A scientist who classifies organisms into groups
Zygote	The splitting of one species into two or more
Taxonomist	Distinct physical characteristics that separate a species
Hybridization	A branch that contains all descendants of a common ancestor
Monophyletic	The first cell that results when a sperm fertilizes an egg

EK 1.C.3
Student Edition p. 297

What is the difference between macro- and microevolution?

_____

_____

_____

A horse and a donkey can mate and produce offspring. Does this mean that they are the same species?

_____

_____

SUMMARIZE IT

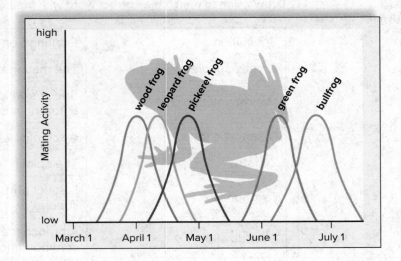

EK 1.C.1, 1.C.3

The figure above illustrates the breeding times of different species of frogs in the genus *Rana*. What type of mechanism may have led to their speciation?

_____

_____

Scientists found another frog living in this region that looked exactly like the leopard frog but had a very different-sounding call.

What would scientists classify this species as?

_____

How would they determine if it was a new species?

_____

_____

What type of event may have led to its speciation, assuming it preferred the same habitat and reproduced at the same time?

_____

Section 17.2   Modes of Speciation, *pp. 303–307*

Essential Knowledge Covered:
1.C.1, 1.C.2

REVIEW IT

EK 1.C.1, 1.C.2
Student Edition pp. 303–306

Identify the mode of speciation given the example below.

Two populations of salmon develop different body size and shape because of geographical isolation	
Cichlid fishes living in open water need to feed on different food then cichlids living on the coastline	
Silversword plants in Hawaii have adapted to live in lava fields as well as dry and wet environments	

Describe convergent evolution.

_____

_____

USE IT

EK 1.C.2
Student Edition p. 305

Imagine a small population of furry creatures colonized a chain of islands, and their descendants spread out to occupy many different niches. How might populations become different as a result of adaptive radiation?

_____

_____

_____

EK 1.C.2
Student Edition p. 305

Create a flow chart to describe how polyploidy in plants can lead to the evolution of a new plant species.

SUMMARIZE IT

EK 1.C.1

Explain how the extinction of the dinosaurs 66 million years ago allowed for the diversification of mammals.

_____

_____

_____

Section 17.3   Principles of Macroevolution, *pp. 308–312*

Essential Knowledge Covered:
1.C.1, 1.C.2, 1.C.3

EK 1.C.1, 1.C.2
Student Edition pp. 308–309

REVIEW IT

Fill in the correct word to complete the following sentences.

The _____ of evolution proposes that speciation occurs after populations become isolated and each group continues on its own evolutionary pathway.

The _____ explains the fluctuating pace of evolution.

Name two genes which have been identified as able to create radical changes in body shapes and organs.

_____

_____

EK 1.C.1, 1.C.3
Student Edition p. 312

Does evolution have a goal?

_____

_____

USE IT

EK 1.C.1
Student Edition p. 310

Which model of evolution is correct, the gradualistic or punctuated equilibrium model?

_____

_____

_____

EK 1.C.1
Student Edition pp. 310–311

Explain how your eyes are related to the eyes of an octopus and the eyes of a fly even though they are very different in their structure and function.

_____

_____

_____

SUMMARIZE IT

EK 1.C.1

Describe how *Hox* genes may have influenced macroevolution.

_____

_____

_____

Section 17.3   Principles of Macroevolution (continued)

EK 1.C.1, 1.C.2, 1.C.3

What does the family tree of the modern horse *Equus* tell us about the evolution of a species?

_____

_____

_____

AP Reviewing the Essential Questions

These questions were posed in the *Biology* chapter opener (page 296). Answer them using the knowledge you've gained from this chapter.

1. What is the role of reproductive isolation in the evolution of new species? How can species maintain reproductive isolation even when occupying a habitat with many other species?

_____

_____

2. How can adaptive radiation accompanied by convergent evolution produce similar assemblages of morphological types in geographically isolated, but similar, environments?

_____

_____

_____

3. How does the universal sharing of some developmental genes in all life forms provide evidence that macroevolution is the source of biodiversity on Earth?

_____

_____

_____

18 Origin and History of Life

As you work through your AP Focus Review Guide, keep this chapter's Big Ideas in mind:

FOLLOWING THE BIG IDEAS

 The theory of evolution does not explain the origin of life but how life on Earth became diverse after life began, and that history can be summarized by macroevolutionary change witnessed in the fossil record. However, early life scientists have developed several hypotheses on possible life origins.

Section 18.1   Origin of Life, *pp. 318–322*
Essential Knowledge Covered:
1.B.1, 1.D.1

EK 1.B.1, 1.D.1
Student Edition pp. 318–322

REVIEW IT

Match the definition to the vocabulary term below.

Term	Definition
abiogenesis	the formation of organic molecules from inorganic material
abiotic synthesis	matter that makes up living organisms
biomolecules	the origin of life from nonliving matter
coacervate droplets	an ancestor common to all organisms that live, and have lived, on Earth
iron-sulfur world	first cell had a plasma membrane before any other parts
last universal common ancestor	complex units of macromolecules
membrane-first hypothesis	dissolved gases from thermal vents and nickel/iron surfaces drove chemical evolution
primordial soup hypothesis	first stage in the origin of life was evolution of simple organic molecules from the Earth's atmosphere
protein-first hypothesis	only RNA needed to form the first cell
proteinoids	organic polymers enclosed in a membrane
protocell	small poly peptide with catalytic properties
RNA-first hypothesis	protein enzymes arose prior to DNA

List the four stages which describe how life originated from nonliving matter.

_____

Section 18.1   Origin of Life (continued)

USE IT

EK 1.D.1
Student Edition p. 321

Draw a diagram to explain how the first plasma membrane might have developed. Include in your answer the words micelle, vesicle, and protocell.

SUMMARIZE IT

EK 1.D.1

Which came first: DNA, RNA, or proteins?

_____

_____

_____

_____

Section 18.2   History of Life, *pp. 323–331*

Essential Knowledge Covered:
1.A.4, 1.B.1, 1.C.3, 1.D.2

REVIEW IT

EK 1.A.4, 1.D.2
Student Edition pp. 323–331

Fill in the blanks to complete the sentences below.

The gradual settling of particles of silt is called _____ .

The _____ protects organisms from harmful ultraviolet rays.

The _____ divides the history of the Earth into eras and then periods and epochs.

Mammalian _____ died out by the end of the Pleistocene.

EK 1.C.3
Student Edition pp. 327–328

Explain the difference between extinction and mass extinction.

_____

_____

_____

Section 18.2   History of Life (continued)

EK 1.A.4, 1.D.2
Student Edition pp. 325–327

List four pieces of modern evidence that supports the endosymbiotic theory.

_____

_____

_____

_____

USE IT

EK 1.B.1, 1.D.2
Student Edition p. 326

Use the flow diagram below to describe in general terms when and how ferns evolved from algae.

SUMMARIZE IT

EK 2.A.2, 1.D.2

Why were ancient cyanobacteria important for the evolution of other living organisms?

_____

_____

EK 1.B.1

What is the Cambrian explosion and why do biologists find it so interesting?

_____

_____

_____

_____

Section 18.3   Geological Factors That Influence Evolution, *pp. 332–334*

Essential Knowledge Covered:
1.C.1

REVIEW IT

Using the vocabulary on the left, **fill in the blanks** below.

Vocabulary

continental drift
plate tectonics

The positions of the continents are not fixed and change over time in a phenomenon

called _____. This phenomenon is studied by a branch of geology known as

_____ which follows the movement of pieces of the Earth's crust which float

on a lower, hot mantle layer.

EK 1.C.1
Student Edition pp. 333–334

USE IT

List the five mass extinctions that have occurred on Earth and give the probable cause of the extinction.

Extinction	Cause

EK 4.B.4

SUMMARIZE IT

How did continental drift lead to differences in the distribution and diversity of marsupials between South America and Australia?

_____

_____

_____

_____

AP Reviewing the Essential Questions

These questions were posed in the *Biology* chapter opener (page 317). Answer them using the knowledge you've gained from this chapter.

1. How does the "organic soup" model support the hypothesis that life could have originated as a result of conditions present on early Earth? What was the role of natural selection in the evolution of cells from organic precursors?

_____

_____

_____

_____

2. How does the fossil record provide evidence for the origin of life on Earth approximately 3.5 billion years ago?

_____

_____

_____

3. What evidence, drawn from many scientific disciplines including geology and molecular biology, supports the hypothesis that all organisms on Earth share a common ancestor?

_____

_____

_____

19 Taxonomy, Systematics, and Phylogeny

As you work through your AP Focus Review Guide, keep this chapter's Big Ideas in mind:

FOLLOWING THE BIG IDEAS

 Macroevoelution explains how new species originate. Systematic biology reconstructs a "tree" of common ancestors and their descendants to investigate the evolutionary relationships among living organisms.

Section 19.1 Systematic Biology, *pp. 338–340*

Essential Knowledge Covered:
1.B.2

EK 1.B.2
Student Edition p. 339

REVIEW IT

List the eight main categories of biological classification.

_____

_____

EK 1.B.2
Student Edition p. 339

USE IT

The specific epithet of a tiger is written below.

Circle the genus and draw a **square** around the species.

> *Panthera   tigris*

EK 1.B.2
Student Edition p. 340

Explain how DNA can be used to identify and classify new species.

_____

_____

_____

EK 1.B.2

SUMMARIZE IT

In 2006, a giant creature that looked like a sea anemone was found at the bottom of the ocean and named *Boloceroides daphneae*. If you look up this organism today you will see that is now called *Relicanthus daphneae*. Why might a scientist change an organism's name?

_____

_____

_____

Section 19.2   The Three-Domain System, *pp. 341–344*

Essential Knowledge Covered:
1.B.2

EK 1.B.2
Student Edition pp. 341–342

USE IT

List two ways how bacteria are different from archaea.

_____

_____

_____

_____

EK 1.B.2
Student Edition p. 344

Some protists and fungi are single-celled organisms. Why are they classified as eukaryotes and not as bacteria or archaea?

_____

_____

SUMMARIZE

EK 1.B.2

Fill in the following table to distinguish the differences between the three domains of life.

Trait	Bacteria	Archaea	Eukarya
Single-celled			
Membrane lipids			
Ribosomes			
Cell Wall			
Nuclear envelope			
Introns			

Section 19.3   Phylogeny, *pp. 344—349*

Essential Knowledge Covered:
1.A.4, 1.B.2

EK 1.B.2
Student Edition pp. 347—348

REVIEW IT

Compare and contrast homology and analogy.

_____

_____

_____

USE IT

EK 1.A.4
Student Edition p. 349

What is a molecular clock?

_____

_____

EK 1.A.4, 1.B.2
Student Edition p. 349

If there are 3 amino acids differences in the protein cytochrome c between horses and pigs, 10 amino acids different between chickens and pigs, and 12 between horses and chickens, what can be concluded about the evolutionary relationships between these organisms?

_____

_____

SUMMARIZE IT

EK 1.B.2

Construct a cladogram, using the chart of physical traits in the organisms listed below.

Trait	Kangaroo	Human	Koala	Cat	Whales
Placental mammal		X		X	X
Fins					X
Four limbs	X	X	X	X	
Canine teeth		X	X	X	
External tail	X			X	X

These questions were posed in the *Biology* chapter opener (page 337). Answer them using the knowledge you've gained from this chapter.

1. What is the connection between the study of macroevolution and the study of systematic biology?

   _____

   _____

   _____

2. What types of scientific data can be used to construct phylogenetic trees or cladograms to visualize the evolutionary history of a group(s) of organisms?

   _____

   _____

   _____

3. How can knowledge of the evolutionary relationships among organisms help address contemporary issues, such as conservation, medicine, and agriculture?

   _____

   _____

   _____

20 Viruses, Bacteria, and Archaea

As you work through your AP Focus Review Guide, keep this chapter's Big Ideas in mind:

FOLLOWING THE BIG IDEAS

BIG IDEA 1
Bacteria are the most ancient form of life on Earth; the origin of viruses remains unknown.

BIG IDEA 2
Prokaryotes include members who use energy in many forms and under diverse conditions.

BIG IDEA 3
Viral and bacterial genomes benefit from their ability to evolve rapidly.

BIG IDEA 4
Prokaryotes can achieve cooperative relationships which benefit all members.

Section 20.1   Viruses, Viroids, and Prions, *pp. 354–360*

Essential Knowledge Covered:
1.C.3, 3.A.1, 3.C.3

Vocabulary

bacteriophages
capsid
emerging viruses
lysogenic cells
lysogenic cycle
lytic cycle
neurodegenerative diseases
prions
retroviruses
reverse transcriptase
viroids
virus

EK 3.C.3
Student Edition p. 356

REVIEW IT

Using the vocabulary on the left, **fill in the blanks** below.

A _____ is known as an obligate intercellular parasite. Viruses are composed of a _____ and an inner core of nucleic acid. Viruses that invade bacteria are known as _____ and alternate between two life cycles: the _____ and the _____. _____ carry prophage genes. In animals, it is possible that viruses can carry RNA, convert it into DNA, and integrate it into the host's genome with help of the enzyme _____. These viruses are known as _____. New viruses that infect large numbers of humans are called _____ and are particular difficult to cure due to their rapid evolution. While not technically viruses because they lack capsids, _____ are able to cause disease in plants. Infectious protein particles, or _____, can cause _____ by forming clusters in the brain.

List the five steps of the reproductive cycle of viruses.

_____

_____

_____

_____

_____

Section 20.1 Viruses, Viroids, and Prions (continued)

EK 3.C.3
Student Edition pp. 356—357

USE IT

How can a virus cause a bacterium that causes strep throat cell to produce a toxin that causes scarlet fever?

_____

_____

_____

EK 3.A.1
Student Edition p. 358

Using a labeled diagram, **explain** how some RNA viruses use reverse transcription in order to be integrated into the host genome to produce more viruses.

SUMMARIZE IT

EK 1.C.3

Describe why H5N1 and H7N9 are considered emerging viruses.

_____

_____

_____

_____

Section 20.2 The Prokaryotes, *pp. 360—363*

Essential Knowledge Covered:
1.B.1, 2.D.2, 3.C.2, 3.C.3

REVIEW IT

Given the definition on the left, **fill in** the correct vocabulary word on the right.

Vocabulary

binary fission
conjugation
conjugation pilus
prokaryotes
transduction
transformation

Definition	Vocabulary Word
When a cell picks up free pieces of DNA secreted or released by a prokaryote	
When two bacteria link together and one passes DNA to the other in the form of a plasmid	
Fully-functioning, living, single-celled organisms	
Asexual reproduction in prokaryotes	
Process in which bacteriophages carry portions of DNA from bacterial cell to another	
The elongated, hollow appendage used to transfer DNA between bacterial cells	

Section 20.2   The Prokaryotes (continued)

EK 1.B.1
Student Edition pp. 360–362

List the three types of appendages a prokaryote might have.

_____

_____

USE IT

EK 2.D.2
Student Edition p. 360

The fossil record suggests that prokaryotes were alone on Earth for 2.5 billion years. How does this information help us better understand the diversity we see in prokaryotes today?

_____

_____

_____

EK 3.C.2
Student Edition p. 363

List three processes prokaryotes can carry out which allow for the acquisition and recombination of new genetic information.

_____

_____

EK 3.C.2
Student Edition p. 363

Why are the three processes you listed above important for increase of new genetic material in prokaryotes?

_____

_____

_____

SUMMARIZE IT

EK 3.C.2, 3.C.3

Compare and contrast transduction and transformation.

Transduction   Transformation

Section 20.3   The Bacteria, *pp. 363–367*

Essential Knowledge Covered:
2.D.1, 2.E.3, 3.D.4, 4.B.2

REVIEW IT

EK 2.D.1
Student Edition pp. 364–365

Identify the bacteria based on the description of its metabolic activity.

Metabolism	Bacteria
Autotrophic bacteria which carry out chemosynthesis	
Heterotrophic bacteria which obtain organic nutrients from other living organisms	
Autotrophic bacteria which are photosynthetic	
Heterotrophic bacteria which break down organic matter from dead organisms	
Gram-negative bacteria that photosynthesize and may contain other pigments as well	

Using the vocabulary on the left, **fill in the blanks** to review some physical characteristics of bacteria.

Vocabulary

bacteria
endospores
peptidoglycan
toxin

The more common type of prokaryotes are _____. They are protected by a

cell wall composed of _____. When faced with unfavorable conditions, some

bacteria can release a portion of their cytoplasm and a copy of their chromosome in

what is known as a(n) _____. Some bacteria can produce a(n) _____,

which is a substance that causes damage to other organisms.

USE IT

EK 2.E.3, 4.B.2
Student Edition p. 365

Describe a mutualistic relationship between bacteria and a eukaryote. Use a specific example you found interesting in the text. Be sure to identify why it's considered mutualistic.

_____

_____

_____

EK 2.D.1, 2.E.3, 4.B.2
Student Edition pp. 365–366

Identify if the relationship is either parasitic (P), commensal (C), or mutualistic (M).

An obligate anaerobe lives in our gut because *E. coli* has used up the oxygen.   ____

A prokaryote digests cellulose in the gut of a goat, and the goat eats grass.   ____

The bacterium, *Clostridium tetani,* enters a cut on someone's foot and colonizes the site of the wound.   ____

A cyanobacterium resides in the nodule of a legume root and the plant is able to obtain nitrogen from it.   ____

A pathogen, *Shigella dysenteriae,* sticks to the human intestinal wall and produces Shiga toxin.   ____

Section 20.3   The Bacteria (continued)

SUMMARIZE IT

EK 2.D.1

Why do some *Salmonella* strains make us sicker than others?

_____

_____

_____

EK 2.D.1

Suppose the water in a lake near a big farm suddenly becomes cloudy and green in the middle of the summer. A scientist on a local news channel says that it is full of cyanobacteria because of high levels of fertilizer present in the area. What does this mean? What does it mean for the future of the lake?

_____

_____

_____

Section 20.4   The Archaea, *pp. 368–369*

Essential Knowledge Covered:
2.A.2, 2.D.2, 4.B.2

REVIEW IT

EK 2.A.2, 2.D.2
Student Edition pp. 368–369

Fill in the chart to describe the three types of archaea.

Name	Habitat	Energy Capturing Mechanism	A Unique Trait
Methanogens			
Halophiles			
Thermoacid-iophiles			

List three reasons why archaea are more similar to eukaryotes then bacteria.

_____

_____

Section 20.4   The Archaea (continued)

USE IT

EK 4.B.2
Student Edition p. 369

Describe how a nitrifying marine archaea may impact other organisms living in the ocean.

_____

_____

_____

EK 2.D.2
Student Edition p. 369

What features allow many archaea to function at high temperatures?

_____

_____

EK 2.D.2
Student Edition p. 369

How have halophiles adapted to living in saline environments?

_____

_____

_____

SUMMARIZE IT

EK 2.A.2

A hydrothermal vent is located in a deep-ocean vent. Here no sunlight shines, the water can reach up to 140°C, and is rich in acidic hydrogen sulfide gas. What type of archaea could a scientist expect to find here and why?

_____

_____

_____

EK 4.B.2

Around the vent there is a complex biological community thriving with microinvertebrates, fish, and octopuses. How do these archaea help in building this community?

_____

_____

_____

These questions were posed in the *Biology* chapter opener (page 353). Answer them using the knowledge you've gained from this chapter.

1. How can studying the evolution of microorganisms help researchers develop vaccines and treatments for disease?

   _____

   _____

   _____

2. How do viruses and prokaryotic bacteria differ from each other and eukaryotes?

   _____

   _____

   _____

3. How can viruses and bacteria introduce the genetic variation necessary for evolution into populations?

   _____

   _____

   _____

4. What is an example of cooperative relationship involving bacteria? What benefits are experienced by the bacteria and other organisms?

   _____

   _____

   _____

21 Protist Evolution and Diversity

As you work through your AP Focus Review Guide, keep this chapter's Big Ideas in mind:

FOLLOWING THE BIG IDEAS

 BIG IDEA 1 The inferred ancestry and diversification of protists can be charted via phylogenetic trees and cladograms.

 BIG IDEA 2 Both biotic and abiotic factors affect the size and activity level of members of the kingdom Protista.

 BIG IDEA 4 Protists may interact with each other or with other species, producing beneficial or harmful consequences.

Section 21.1 General Biology of Protists, *p. 374*

PREREQUISITE KNOWLEDGE

This section covers concepts you should know from other science courses. Review it to master the AP Essential Knowledge and prepare for the AP Exam.

REVIEW IT

Student Edition p. 374

Match the vocabulary word with its definition.

cysts	simplest but most diverse eukaryotes
endosymbiosis	from a single evolutionary lineage
monophyletic	acquiring organelles by engulfing free-living bacteria
mixotrophic	suspended aquatic organisms that are an important food source
plankton	a combination of auto- and heterotrophic nutritional modes
protists	dormant life cycle phase
supergroup	high level taxonomic group

Student Edition p. 374

How are protists categorized?

_____

_____

_____

USE IT

Student Edition p. 374

How do protists reproduce?

_____

_____

Section 21.1   General Biology of Protists (continued)

SUMMARIZE IT
Describe two ways in which protists play a significant ecological role in aquatic systems.

_____

_____

_____

_____

Section 21.2   Supergroup Archaeplastida, *pp. 377–379*

Essential Knowledge Covered:
2.D.1, 2.D.2

REVIEW IT
Given the definition on the left, **fill in** the correct vocabulary word on the right.

Vocabulary

archaeplastids
charophytes
colony
green algae
filaments
red algae
zoospores

Definition	Vocabulary Word
The green algae group most closely related to land plants	
A supergroup that contains land plants and other photosynthetic organisms which have plastids derived from endosymbiotic cyanobacteria	
Protist which contain both chlorophylls *a* and *b*	
Multicellular seaweeds that possess red and blue accessory pigments	
End-to-end chains of cells	
Haploid flagellated spores that grow to become adult vegetative cells	
A loose association of independent cells	

USE IT

EK 2.D.1
Student Edition p. 379

List two ways in which red algae is economically important.

_____

_____

_____

SUMMARIZE IT

EK 2.D.1, 2.D.2

How do environmental conditions effect the reproductive mechanisms of *Chlamydomonas*?

_____

_____

_____

_____

Vocabulary

apicomplexans
brown algae
cillates
diatoms
dinoflagellates
golden brown algae
water molds

REVIEW IT

Identify the chromalveolate based on its description.

Description	Chromalveolate
Stramenopiles with chlorophyll *a*, *c* and a brown-green accessory pigment	
Single-celled alveolates that move by means of cilia	
Fungus-like stramenopiles that caused potato famine in the 1840s	
Single-celled alveolates encased by protective cellulose and silicate plates	
Stramenopiles with chlorophyll *a* and *c* and a yellow-green accessory pigment	
Alveolates with a unique organelle called an apicoplast which can penetrate a host cell	
Single celled stramenopiles with an ornate silica shell	

USE IT

EK 2.D.2
Student Edition p. 380

Kelp and *Fucus* are brown algae that live along the shoreline in temperate regions. What sorts of mechanisms have they evolved to be able to withstand both powerful waves and exposure to air?

_____

_____

K 2.D.1
Student Edition p. 382

What are red tides, and how can they be dangerous?

_____

_____

_____

SUMMARIZE IT

EK 2.D.1, 2.D.2, 4.B.4, 4.C.3

Describe what causes malaria and how it is spread through a population.

_____

_____

_____

_____

Section 21.4   Supergroup Excavata, *pp. 385–389*

Essential Knowledge Covered:
2.D.1, 2.D.2, 4.B.4, 4.C.3

Vocabulary

diplomanad
euglenids
kinetoplastids
parabasalids

REVIEW IT

Identify the excavata based on its description.

Description	Excavata
Single-celled, flagellated protozoans with large masses of DNA in their mitochondria	
A single-celled excavate with two nuclei and two sets of flagella	
A single-celled, flagellated excavate with a unique, fibrous connection between the Golgi apparatus	
A small, single celled freshwater excavate that produces an unusual type of carbohydrate	

USE IT

EK 2.D.1, 2.D.2
Student Edition p. 385

Why is it difficult to classify the euglenids?

_____

_____

_____

EK 2.D.1, 2.D.2, 4.B.4, 4.C.3
Student Edition pp. 388–389

Draw a diagram of how African Sleeping Sickness is transmitted.

SUMMARIZE IT

EK 2.D.1, 4.B.4, 4.C.3

Describe a way in which climate change may increase the rate of protist-related human disease.

_____

_____

_____

_____

Section 21.5 Supergroups Amoebozoa, Opisthokonta, and Rhizaria, *pp. 389–391*

Essential Knowledge Covered:
1.B.1, 2.D.2

EK 2.D.1
Student Edition p. 389

REVIEW IT

Describe how amoebozoans move.

_____

_____

EK 1.B.1, 2.D.1
Student Edition p. 390

USE IT

In what ways do choanoflagellates resemble sponges?

_____

_____

EK 4.B.4

SUMMARIZE IT

How do foraminiferans help shape the environment?

_____

_____

_____

_____

AP Reviewing the Essential Questions

These questions were posed in the *Biology* chapter opener (page 373). Answer them using the knowledge you've gained from this chapter.

1. What explains the amazing diversity of protists, and why is it challenging to classify them?

_____

_____

_____

2. How do protists affect other organisms and the environment?

_____

_____

_____

3. How do microorganisms, including protists, impact our health and welfare?

_____

_____

22 Fungi Evolution and Diversity

As you work through your AP Focus Review Guide, keep this chapter's Big Ideas in mind:

FOLLOWING THE BIG IDEAS

BIG IDEA 1 The inferred ancestry and diversification of fungi can be charted via phylogenetic trees and cladograms.

BIG IDEA 2 The environment exerts influence of many aspects of fungi physiology and reproduction.

BIG IDEA 4 Fungi engage in both beneficial and harmful relationships with other organisms.

Section 22.1   Evolution and Characteristics of Fungi, *pp. 396–397*

Essential Knowledge Covered:
1.B.1, 2.C.2

Vocabulary

asptate
budding
dikaryotic
chitin
fungi
hyphae
mycelium
saprotrophs
septa
septate
spores

REVIEW IT

Using the vocabulary on the left, **fill in the blanks** below.

_____ are heterotrophs that are spilt into six groups. As they obtain their nutrient by absorbing food, fungi are known as _____. Some fungi produce filaments called _____ which grow from their tips and may have cross-walls known as _____. Fungi with septa are called _____ and _____ if the septa do not have cross-walls. A hypha that contains paired haploid nuclei is called _____. Hyphae may form a network called a _____. Fungi cells are quite different from plant cells because they contain _____. Fungi usually produce _____ during sexual and asexual reproduction. Other fungi can reproduce asexually by _____.

USE IT

EK 2.C.2
Student Edition p. 397

How are fungi able to disperse their offspring?

_____

_____

_____

EK 1.B.1, 2.C.2
Student Edition p. 397

Describe two ways fungi are more like animals than plants.

_____

_____

Section 22.1 Evolution and Characteristics of Fungi (continued)

SUMMARIZE IT

EK 2.C.2

Draw a picture showing the three stages of sexual reproduction in terrestrial fungi.

Section 22.2 Diversity of Fungi, *pp. 398—405*

Essential Knowledge Covered:
2.C.2, 2.E.2, 2.E.3, 4.B.3,
4.C.3

REVIEW IT

Given the definition on the left, **fill in** the correct vocabulary word(s) on the right.

Vocabulary

ascus
AM fungi
basidium
chytrids
club fungi
conidiospores
fruiting body
microsporidia
molds
sac fungi
sporangium
yeasts
zoospores
zygospore

Definition	Vocabulary Word
A capsule that produces haploid spores during asexual reproduction	
A structure that produces spores and enhances their dispersal	
The Glomeromycota	
A club-shaped structure in which spore develop	
The Ascomycota	
A spore produced at the tip of aerial hyphae	
The Zygomycota	
A fingerlike sac that develops duing sexual reproduction	
The Basidiomycota	
Spore with flagella	
A zygote contained in a thick wall	
Single-celled parasitic fungi	
The Chytridiomycota	
Two of the most common morphological types of sac fungi	

Section 22.2 Diversity of Fungi (continued)

USE IT

EK 2.E.2, 2.E.3
Student Edition p. 402

Provide two examples of how different species of club fungi disperse their spores.

_____

_____

_____

EK 2.C.2, 2.E.3
Student Edition pp. 399–400

How does black bread mold survive when conditions are unfavorable?

_____

_____

EK 4.C.3
Student Edition p. 405

Compare and contrast *Geomyces destructans* and *Batrochochytrium dendrobatidis*.

_____

_____

_____

EK 4.B.3
Student Edition p. 401

What type of fungi do leaf cutter ants and elm trees have in common?

_____

_____

_____

SUMMARIZE IT

EK 2.E.2

Draw a picture of the life cycle of black bread mold. When are the cells haploid? When are they diploid?

EK 2.E.2, 2.E.3

Describe two ways the life cycle of bread mold different than that of a typical mushroom.

_____

_____

_____

Section 22.3   Symbiotic Relationships of Fungi, *pp. 405–407*

Essential Knowledge Covered:
2.E.2

EK 2.E.2
Student Edition pp. 405–407

REVIEW IT

Identify the symbiotic relationship below.

The association between a fungus and a cyanobacterium or green alga is a _____.

The association between a soil fungi and a root of a plant is called _____.

USE IT

EK 2.E.2
Student Edition p. 405

Describe the three types of lichen.

_____

_____

EK 2.E.3
Student Edition p. 406

What is the difference between ectomycorrhizae and endomycorrhizae?

_____

_____

SUMMARIZE IT

EK 2.E.2

Why is the formation of lichen no longer considered mutualistic?

_____

_____

_____

_____

EK 2.E.2

Describe a benefit received by the plant and by the fungus in a mycorrhizae relationship.

_____

_____

_____

_____

AP Reviewing the Essential Questions

These questions were posed in the *Biology* chapter opener (page 395). Answer them using the knowledge you've gained from this chapter.

1. What is the evolutionary history of fungi? Are they more closely related to plants of animals?

   _____

   _____

   _____

2. How would ecosystems be impacted if fungi were to go extinct?

   _____

   _____

   _____

3. What properties of fungi make them an important resource for the study of medicine, genetics, and molecular biology?

   _____

   _____

   _____

23 Plant Evolution and Diversity

As you work through your AP Focus Review Guide, keep this chapter's Big Ideas in mind:

FOLLOWING THE BIG IDEAS

 BIG IDEA 1 Plants evolved from aquatic ancestors, eventually adapting to land with developments of vascular tissues, fertilization without water, and seeds.

Section 23.1 Ancestry and Features of Land Plants, *pp. 412–415*

Essential Knowledge Covered:
1.A.1, 1.B.2

EK 1.B.2
Student Edition p. 412

REVIEW IT

List two ways green algae are like plants and one major difference.

_____

_____

USE IT

EK 1.A.1
Student Edition p. 414

Draw and explain a diagram of the alternation of generations in land plants. Label the diploid and haploid structures as well when meiosis and mitosis occur.

EK 1.B.2
Student Edition p. 412

According to phylogenetic information, in what order did modern plant groups diverge?

_____

_____

SUMMARIZE IT

EK 1.A.1

Describe four adaptations that allowed plants to move to land.

_____

_____

_____

_____

Section 23.2 Evolution of Bryophytes: Colonization of Land, *pp. 415—417*

Essential Knowledge Covered:
1.B.2

EK 1.B.2
Student Edition p. 415—417

REVIEW IT

List two defining characterizes for each bryophyte.

Moss	Liverwort	Hornwort

SUMMARIZE IT

EK 1.B.2

Compare and contrasts traits of bryophytes with traits of vascular plants.

_____

_____

_____

_____

Section 23.3 Evolution of Lycophytes: Vascular Tissue, *pp. 417—419*

Essential Knowledge Covered:
1.A.1, 1.B.1

USE IT

EK 1.B.1
Student Edition p. 419

Describe the difference between a microspore and a megaspore.

_____

_____

EK 1.B.1
Student Edition p. 419

List two ways the branching of *Cooksonia* supports important information concerning the evolution of vascular plants.

_____

_____

SUMMARIZE IT

EK 1.A.1, 1.B.1

How was xylem essential to the evolution of upright and taller plants?

_____

_____

_____

_____

Section 23.4   Evolution of Pteridophytes: Megaphylls, *pp. 419—423*

Essential Knowledge Covered:
1.A.1, 1.B.1, 1.B.2

EK 1.B.2
Student Edition pp. 419–423

REVIEW IT

Identify the pteridophyte based on its description.

Description	Pteridophyte
Members of the genera *Psilotum* or *Tmesipteris*	
All members of the genus *Equisetum*	
The most diverse group of pteridophytes that range in size from 1 cm in diameter to 20 m high	

USE IT

EK 1.B.1
Student Edition p. 420

Describe the difference between megaphyll and a microphyll.

_____

_____

_____

EK 1.A.1, 1.B.1
Student Edition p. 420

List two ways the wide fronds of the fern supports important information concerning the evolution of vascular plants.

_____

_____

SUMMARIZE IT

EK 1.A.1, 1.B.1

Describe two modern day uses for pteridophytes.

_____

_____

_____

> ## EXTENDING KNOWLEDGE
>
> This section takes the AP Essential Knowledge you have learned further, and may provide illustrative examples useful for the AP Exam.

USE IT

Student Edition p. 423

Why was pollen an important evolutionary advantage?

_____

_____

_____

SUMMARIZE IT

Determine if the following statement applies to angiosperms, gymnosperms or both.

Statement		Seed Plant
Can be moncot or eudicot		
Contains conifers, cycads, ginkgoes, and gnetophytes		
Has seeds		
Ovules are completely enclosed and become fruit		
May be monoecious or dioceious		
Ovules are not completely enclosed by sporophyte tissue		

Reviewing the Essential Questions

These questions were posed in the *Biology* chapter opener (page 411). Answer them using the knowledge you've gained from this chapter.

1. What environmental challenges did plants face in their evolution from aquatic to terrestrial environments? What adaptations enabled plants to make this transition?

   _____

   _____

   _____

2. How have humans manipulated plants to better serve our needs?

   _____

   _____

   _____

24 Flowering Plants: Structure and Organization

As you work through your AP Focus Review Guide, keep this chapter's Big Ideas in mind:

FOLLOWING THE BIG IDEAS

 BIG IDEA 2 Important adaptations, including increased surface area and specialized exterior surface, benefit terrestrial plants.

 BIG IDEA 4 The environment affects all aspects of a plant's life; adaptations allow them to be successful in dramatically varying climates.

Section 24.1   Cells and Tissues of Flowering Plants, *pp. 436–439*

Essential Knowledge Covered:
2.A.3, 4.A.4, 4.B.2

Vocabulary

collenchyma
cuticle
meristem
parenchyma
periderm
phloem
root hairs
sclerenchyma
stomata
trichomes
xylem

REVIEW IT

Identify the structure given its function.

Function	Structure
Minimizes water loss and protects against disease	
Transports sucrose and organic compounds from the leaves to the roots	
Open and close for gas exchange and water loss	
Cells which give flexible support to immature regions of the body	
Transports water and minerals from the root to the leaves	
Can assume many different shapes and sizes as they grow	
The cork area of a plant which helps a plant resist predation	
The most abundant cells in a plant which may contain chloroplasts or may just store carbohydrates	
Thick secondary cell walls impregnated with lignin that make cells strong and hard	
Hairs which protect the plant from sun and moisture loss	
Increases the surface area for nutrient and water absorption in roots	

Section 24.1 Cells and Tissues of Flowering Plants (continued)

USE IT

EK 4.A.4, 4.B.2
Student Edition p. 439

Describe how xylem moves water from the roots to the leaves.

_____

_____

_____

EK 4.A.4, 4.B.2
Student Edition p. 439

Describe how phloem moves sucrose from the leaves to the roots.

_____

_____

_____

SUMMARIZE IT

EK 2.A.3, 4.B.2

Describe two modification of epidermal tissue in plants which allow for more efficient use of energy and matter.

_____

_____

_____

Section 24.2 Organs of Flowering Plant, *pp. 440–441*

Essential Knowledge Covered:
2.E.2, 4.A.4

USE IT

EK 4.A.4
Student Edition p. 440

Draw a picture of how a basic plant body is organized. Be sure to identify the root and shoot system. Include in your diagram a terminal bud, stem, node, internode, leaf, and root.

SUMMARIZE IT

EK 4.A.4

How do the root and shoot system in plants interact?

_____

_____

Section 24.2   Organs of Flowering Plant (continued)

EK 2.E.2, 4.A.4

What determines if a plant is an annual or a perennial?

_____

_____

_____

Section 24.3   Organization and Diversity of Roots, *pp. 442–444*

Essential Knowledge Covered:
4.A.6

REVIEW IT

EK 4.A.6
Student Edition pp. 442–444

Define the specialized tissues found in the zone of maturation and place them in order from outermost layer (1) to innermost layer (5).

Tissue	Definition	Order
Cortex		
Pericycle		
Vascular tissue		
Endodermis		
Epidermis		

EK 4.A.6
Student Edition p. 444

Identify the root structure based on its function.

Structure	Function
	Houses nitrogen-fixing bacteria
	Prevents the passage of water and mineral ions between cell walls
	Protects the root apical meristem

USE IT

EK 4.A.6
Student Edition p. 444

Compare and contrast taproots and fibrous roots.

_____

_____

_____

_____

Section 24.3 Organization and Diversity of Roots (continued)

EK 4.A.6

SUMMARIZE IT

Describe two specialized root adaptations found in different plants.

_____

_____

_____

_____

Section 24.4 Organization and Diversity of Stems, *pp. 445–449*

Essential Knowledge Covered:
2.D.3, 2.D.4, 4.B.2

EK 4.B.2
Student Edition pp. 446–448

REVIEW IT

Identify the location of the annual ring, cork, cork cambium, vascular cambium, and wood.

EK 2.D.3, 2.D.4

SUMMARIZE IT

Describe the advantages and disadvantages of being a woody plant.

_____

_____

_____

_____

_____

Section 24.5 Organization and Diversity of Leaves, *pp. 450–453*

Essential Knowledge Covered:
2.A.3, 2.C.1

EK 2.A.3
Student Edition pp. 450–451

USE IT

While leaves can be incredibly diverse in color, shape, size, and placement on a stem, what are some conserved characteristics which all leaves share?

_____

_____

_____

SUMMARIZE IT

EK 2.A.3, 2.C.1

How have the Venus fly trap and sundew plants evolved to live in nitrogen-lacking soils?

_____

_____

_____

_____

AP Reviewing the Essential Questions

These questions were posed in the *Biology* chapter opener (page 435). Answer them using the knowledge you've gained from this chapter.

1. What structural and physiological features of angiosperms have enabled them to become the dominant form of plant life on Earth today?

_____

_____

_____

2. How have humans come to rely on angiosperms for a number of uses?

_____

_____

_____

25 Flowering Plants: Nutrition and Transport

As you work through your AP Focus Review Guide, keep this chapter's Big Ideas in mind:

FOLLOWING THE BIG IDEAS

BIG IDEA 1
All plants share certain common features and processes that reflect common ancestry.

BIG IDEA 2
Acquisition of water and nutrients by plants involves specialized mechanisms and structures.

BIG IDEA 4
Coordination among plant structures and interactions between plants and other organisms allow for the exchange of materials with environment.

Section 25.1 Plant Nutrition and Soil, *pp. 457–460*

Essential Knowledge Covered:
2.A.3, 2.B.2

REVIEW IT

EK 2.A.3
Student Edition pp. 457–460

Match the vocabulary word to its definition.

soil	When water or wind carries soil away to a new location
soil erosion	An inorganic substance with two or more elements
essential nutrients	A mixture of minerals, organics, living organisms, air, and water
mineral	Decaying organic matter
humus	C, H, O, P, K, N, S, Ca, Mg, Fe, B, Mn, Cu, Zn, Mo, Cl
cation exchange	When H^+ switch places with a positively charged mineral ion

EK 2.A.3
Student Edition p. 457

Describe the difference between soil profile and soil horizon.

_____

_____

_____

Section 25.1   Plant Nutrition and Soil (continued)

USE IT

EK 2.B.2
Student Edition p. 457

What makes a nutrient essential?

_____

_____

EK 2.A.3
Student Edition pp. 459–460

List three components of healthy soil.

_____

_____

_____

EK 2.A.3
Student Edition p. 460

Identify two ways soil erosion causes problems.

_____

_____

SUMMARIZE IT

EK 2.A.3

A ratio of humus and clay is important to support healthy plant life in soil. Why is this?

_____

_____

_____

Section 25.2   Water and Mineral Uptake, *pp. 461–464*

Essential Knowledge Covered:
2.D.1, 2.E.3, 4.B.2

REVIEW IT

EK 4.B.2
Student Edition p. 463

What are rhibozia and where do they live?

_____

_____

EK 2.D.1
Student Edition pp. 462–464

Determine if the following statements about mycorrhizae are true or false (T or F).

Mycorrhizare are bacteria which live on plant roots _____

Most all plants have mycorrhizae _____

Mycorrhizae increase the surface area for mineral and water uptake _____

Plants do not depend on mycorrhizae _____

Section 25.2   Water and Mineral Uptake (continued)

USE IT

EK 2.D.1
Student Edition pp. 462–464

Compare and contrast mycorrhizae and rhizobia.

_____

_____

_____

EK 2.D.1
Student Edition p. 461

List the two ways water can enter a root.

_____

_____

SUMMARIZE

EK 2.E.3, 4.B.2

Why do plants grow better when mycorrhizae are present?

_____

_____

_____

EK 4.B.2

Plants do not always obtain nutrients through mutualistic means. **Describe** two adaptations in plants that have evolved to obtain nutrients through other means.

_____

_____

Section 25.3   Transport Mechanisms in Plants, *pp. 465–473*

Essential Knowledge Covered:
1.B.1, 2.C.1, 2.E.2, 2.E.3, 4.B.2

REVIEW IT

EK 2.C.1
Student Edition p. 471

Describe the relationship between source and sink.

_____

EK 1.B.1, 2.C.1
Student Edition pp. 465–470

Identify the process based on the description provided.

Dew drops form on leaves	
Water molecules cling together and form a chain	
Water enters the root	
Water molecules stick to xylem	
Water moves from one place to another	
Water and minerals travel up xylem	

Section 25.3   Transport Mechanisms in Plants (continued)

EK 2.C.1, 4.B.2

USE IT

Explain the process occurring in the following diagram.

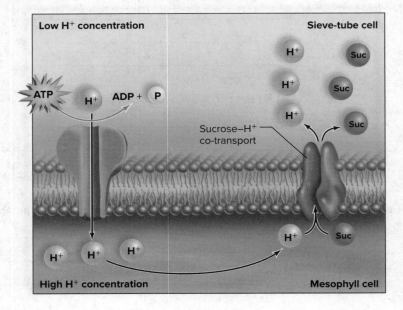

_____

_____

_____

_____

EK 2.C.1, 2.E.2, 2.E.3
Student Edition p. 470

Determine if the following will make a stomata open (O) or close (C).

External carbon dioxide levels increase ____

Daylight shines on the stomata ____

Potassium ions enter the guard cells ____

Guard cells take up water ____

Abscisic acid is produced by wilting leaves ____

SUMMARIZE IT

EK 2.C.1, 2.E.2

Explain how water moves from the soil to the roots to the shoots and out of the leaves of a plant through changes in water potential and through transpiration.

_____

_____

_____

_____

AP Reviewing the Essential Questions

These questions were posed in the *Biology* chapter opener (page 456). Answer them using the knowledge you've gained from this chapter.

1. How do the properties of water, especially adhesion and cohesion, facilitate the transport of water from the roots to the leaves?

   _____

   _____

   _____

2. How do environmental factors influence the opening and closing of stomata and, consequently, the rate of transpiration?

   _____

   _____

   _____

3. How do symbiotic relationships assist plants in acquiring nutrients from the soil?

   _____

   _____

   _____

26 Flowering Plants: Control of Growth Responses

As you work through your AP Focus Review Guide, keep this chapter's Big Ideas in mind:

FOLLOWING THE BIG IDEAS

 BIG IDEA 2   Growth and timing responses are essential to plant energy acquisition and survival.

 BIG IDEA 3   Plant hormones typically work by affecting gene expression in some manner.

Section 26.1   Plant Hormones, *pp. 477–484*

Essential Knowledge Covered:
2.C.1, 2.D.4, 2.E.2, 3.B.2,
3.D.1, 3.D.3

EK 3.B.2
Student Edition pp. 477–480

REVIEW IT

Identify the plant hormone that triggers the response.

Increases the length of a stem _____

Causes a plant to grow taller and taller _____

Promotes cell division _____

Closes stomata _____

USE IT

EK 3.D.3
Student Edition p. 479

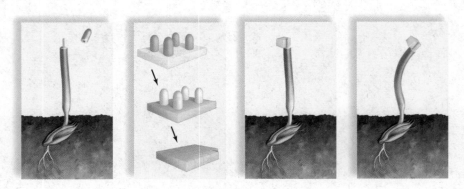

The figure above illustrates an important study that determined how phototrophism occurs in seedlings. Explain what is happening in the illustration and the conclusion of the study.

_____

_____

_____

_____

Section 26.1 Plant Hormones (continued)

SUMMARIZE IT

EK 2.C.1, 3.B.2, 3.D.3

Imagine you are a grocery store owner, and you are shipped green tomatoes.
How is it possible to ripen the tomatoes even though they are already off the vine?

_____

_____

_____

EK 2.E.2, 2.D.4, 3.B.2, 3.D.1

Two insects begin to feed on a plant leaf. The first insect takes a bite without any
problem. The second insect takes a bite and experiences a terrible sensation. Why did
the first insect have no problem eating the leaf but the second insect does?

_____

_____

_____

_____

_____

Section 26.2 Plant Growth and Movement Reponses, *pp. 485–489*

Essential Knowledge Covered:
2.C.2, 2.E.2

REVIEW IT

EK 2.C.2
Student Edition pp. 485–486

Identify the plant movement.

A plant grows towards light	
A plant grows toward or away from a stimuli	
A plant moves because of changes in water potential inside	
A plant moves in response to being touched	
A plant grows against gravity	

USE IT

EK 2.E.2
Student Edition pp. 485–486

Explain the difference between a positive and negative tropism and give an example
of each.

_____

_____

_____

_____

Section 26.2 Plant Growth and Movement Reponses (continued)

EK 2.C.2, 2.E.2
Student Edition pp. 488–489

Using the words *increase* or *decrease*, **fill in the blanks** to describe why a plant bends towards light.

When sun shines on a plant, there is an _____ in auxins in the shaded cells. This leads to an _____ of hydrogen ions, and a _____ in the pH. The internal acidic environment causes a _____ in the integrity of the cell walls. Responding to the _____ in turgor pressure, water from other parts of the plant flow into these cells. There is an _____ in the length of the cells as the plant builds new walls and the turgor pressure is restored. The _____ in size of the cells on the shady side causes the plant to bend toward the light.

SUMMARIZE IT

EK 2.C.2, 2.E.2

How does phototropism occur in plants?

_____

_____

_____

_____

_____

EK 2.E.2

The sensitive plant, *Mimosa pudica*, will collapse its whole leaf within a second after you touch it. **Explain** how it is able to do this.

_____

_____

_____

Section 26.3 Plant Responses to Phytochrome, *pp. 489–493*

Essential Knowledge Covered:
2.C.2, 2.E.2

Vocabulary

biological clock
circadian rhythm
day-neutral plants
etiolated
long-day plants
photoperiodism
phytochrome
short-day plants

REVIEW IT

Using the vocabulary on the left, **fill in the blanks** below.

A _____ refers to a cycle of activity in organism during a 24 hour period. The mechanism that keep the circadian rhythm maintained in the absence of sunlight is an example of a _____. A physiological response that is dependent on the length of day or night is called _____. _____ require a longer period of dark than light in order to flower. _____ require a shorter period of dark than light to flower. Plants that are not dependent on day length for flowering are called _____. The activity of the leaf pigment _____ is necessary for photoperiodism to occur. Plants grown in the dark tend to be _____, which means their stems are elongated and their leaves are small and yellow.

USE IT

EK 2.E.2
Student Edition p. 490

A scientist is trying to classify a plant as a short-day or long-day plant. If the plant still blooms even when the night cycle is interrupted with a bright flash of light, what type of plant is it?

_____

_____

EK 2.C.2, 2.E.2
Student Edition p. 489

List two ways in which plants respond to changes in light.

_____

_____

_____

_____

SUMMARIZE IT

EK 2.C.2, 2.E.2

Suppose you pick up a packet of seeds. On the back, it says to only partly cover the seeds with soil when planting and then space the seedling when they germinate 10-12 inches apart. Explain the scientific reasoning behind these instuctions.

_____

_____

_____

_____

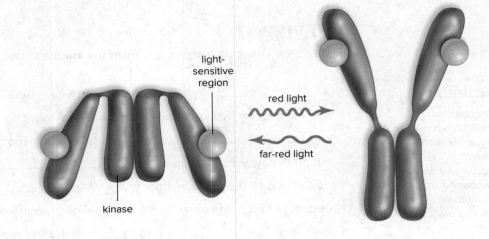

light-sensitive region

kinase

red light

far-red light

EK 2.C.2, 2.E.2
Student Edition p. 490

The figure above shows the conversion of phytochrome from the inactive form to the active form. **Circle** the mechanism which initiates the change and **explain** two functions of the active phytochrome.

_____

_____

_____

AP Reviewing the Essential Questions

These questions were posed in the _Biology_ chapter opener (page 476). Answer them using the knowledge you've gained from this chapter.

1. What are examples of environmental stimuli that trigger various plant responses?

   _____

   _____

   _____

2. What are examples of defenses that plants have evolved that act as deterrents to predation, invasion, and competition?

   _____

   _____

   _____

3. How do plants use signal transduction pathways to respond to environmental stimuli?

   _____

   _____

   _____

27 Flowering Plants: Reproduction

As you work through your AP Focus Review Guide, keep this chapter's Big Ideas in mind:

FOLLOWING THE BIG IDEAS

 Angiosperm reproductive strategies include specialized processes and responses and cooperation with other species.

 Plant reproduction is directly affected by environmental factors.

Section 27.1 Sexual Reproductive Strategies, *pp. 496–501*

Essential Knowledge Covered:
2.E.1, 4.A.6, 4.B.3

Vocabulary

coevolved
double fertilization
endosperm
embryo sac
flowers
pollen grain
pollination
stigma

REVIEW IT

Using the vocabulary on the left, **fill in the blanks** below.

Microspores and megaspores are produced in the reproductive structure known as

_____ in angiosperms. Flowers have several parts but the _____

and _____ are the two structures which are considered the female and male

gametophytes. When pollen is transferred from the anther to the _____,

_____ occurs. Sperm cells travel to the embryo sac and the flowering plant

undergoes _____, where one sperm nucleus unties with the egg nucleus,

forming a 2n zygote, and another forms a 3n _____ cell by uniting with the

polar nuclei. Many flowers and animal pollinators have _____, increasing the

success of flowering plants on land.

EK 2.E.1
Student Edition p. 496

Draw a diagram of a complete flower. Include the following structures: a stamen with anther and filament, a carpel with stigma, style, ovary, and ovule, petals, and sepals.

EK 2.E.1
Student Edition p. 496

USE IT

Identify the following structures in the diagram below:
Seed, pollen grain, microspore, ovule, megaspore, ovary, anther, sporophyte, embryo sac, seed.

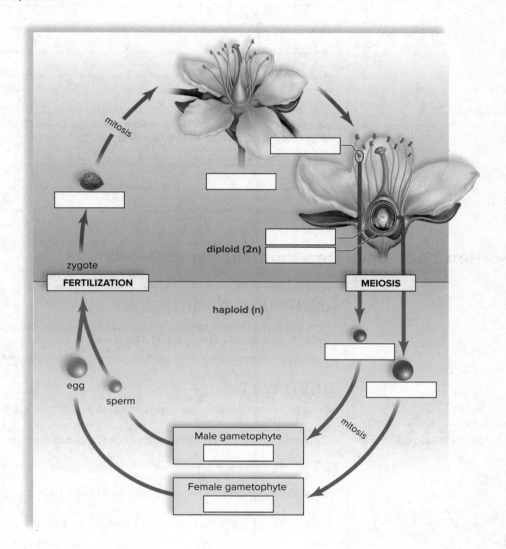

EK 2.E.1
Student Edition pp. 497–499

Describe the events that must occur before a pollen grain can become a mature male gametophyte.

_____

_____

Section 27.1   Sexual Reproductive Strategies (continued)

SUMMARIZE IT

EK 4.B.3

Describe an example of coevolution between an angiosperm and a pollinator that you read about in this section.

_____

_____

_____

_____

EK 4.A.6

Your friend is growing plants that produce yellow, blue, and white flowers, which open during the day. Your friend claims they are bat-pollinated. Is your friend correct? What do you think pollinates these flowers?

_____

_____

_____

Section 27.2   Seed Development, *pp. 502–503*

EXTENDING KNOWLEDGE

This section takes the AP Essential Knowledge you have learned further, and may provide illustrative examples useful for the AP Exam.

REVIEW IT

Identify the vocabulary word based on the definition below.

Student Edition pp. 502–503

The programmed series of stages from simple to a more complex form is

called: _____

The specialization of structures and function is called: _____

The difference between monocots and eudicots is the number of: _____

USE IT

Student Edition pp. 502–503

List and describe three parts of a seed.

_____

_____

_____

SUMMARIZE IT

What is the importance of the seed coat?

_____

_____

Section 27.3   Fruit Types and Seed Dispersal, *pp. 504–506*

REVIEW IT

Student Edition p. 504

Define *fruit*.

_____

_____

USE IT

Student Edition p. 504

Describe two types of fruit dispersal.

_____

_____

_____

Student Edition pp. 504–506

List three requirements for seed germination.

_____

_____

_____

SUMMARIZE IT

What is your favorite fruit? **Classify** your favorite fruit based on its composition.

_____

_____

_____

Section 27.4   Asexual Reproductive Strategies, *pp. 507–509*

Essential Knowledge Covered:
2.C.2, 4.B.2

Vocabulary

asexual reproduction
cell suspension culture
clone
tissue culture
totipotent

EK 4.B.2
Student Edition p. 509

EK 2.C.2

REVIEW IT

Using the vocabulary on the left, **fill in the blanks** below.

Some plants can produce offspring from a single parent through _____.
The genetically identical offspring is known as a _____. In a laboratory, plant
clones are often made through _____. Plant cells are _____,
meaning a single cell can develop into an entire plant. Scientists employ another
technique called _____ to extract chemicals from plant cells to be
used for medicinal purposes.

USE IT

Compare and contrast stolon and rhizomes.

_____

_____

_____

_____

SUMMARIZE IT

Provide one advantage and one disadvantage for asexual reproduction in plants.

_____

_____

_____

_____

These questions were posed in the *Biology* chapter opener (page 495). Answer them using the knowledge you've gained from this chapter.

1. How are the reproductive strategies of flowering plants adapted to a terrestrial lifestyle?

   _____

   _____

   _____

   _____

2. How do interactions among flowering plants, pollinators, and the environment contribute to the success and diversity of the angiosperms?

   _____

   _____

   _____

   _____

28 Invertebrate Evolution

As you work through your AP Focus Review Guide, keep this chapter's Big Ideas in mind:

FOLLOWING THE BIG IDEAS

 Invertebrate evolutionary trends include bilateral symmetry, cephalization, and segmentation.

 Animal body systems help maintain homeostasis but are tailored to specific needs and environments.

Section 28.1   Evolution of Animals, *pp. 514–517*

Essential Knowledge Covered:
1.B.1, 2.E.1

Vocabulary

asymmetry
bilateral symmetry
blastula
cephalization
choanoflagellates
cleavage
coelom
deuterostome
germ layers
invertebrates
protostomes
radial symmetry
symmetry
vertebrates

REVIEW IT

Fill in the vocabulary word or its definition.

Vocabulary Word	Definition
choanoflagellates	
invertebrates	
cephalization	
	The first developmental event after fertilization
	The anus develops prior to the mouth
radial symmetry	
protostomes	
	Defined right and left halves
coelom	
	Animals that have a spinal cord at some stage of their lives
asymmetry	
germ layers	
	A hollow sphere of cells
	A pattern of similarity observed in living organisms

Section 28.1 Evolution of Animals (continued)

USE IT

EK 1.B.1, 2.E.1
Student Edition p. 517

List the three major events of animal development.

_____

_____

EK 2.E.1
Student Edition p. 517

Describe the differences between protostome and deuterostome development and give an example animal for each.

_____

_____

_____

_____

SUMMARIZE IT

EK 1.B.1

After the embryo of an animal with bilateral symmetry develops an anterior and posterior region, describe what genes determine the late of that animal's segmentation pattern.

_____

_____

_____

Section 28.2 The Simplest Invertebrates, *pp. 521–523*

> ### EXTENDING KNOWLEDGE
>
> This section takes the AP Essential Knowledge you have learned further, and may provide illustrative examples useful for the AP Exam.

REVIEW IT

Student Edition p. 521

Identify the structure as either belonging to a Sponge (S), Comb Jelly (CJ) or Cnidarian (C).

nematocyst _____

spicule _____

hydrostatic skeleton _____

mesoglea _____

gastrovascular cavity _____

Section 28.2   The Simplest Invertebrates (continued)

USE IT

Student Edition pp. 521–522

Compare the two body forms of a cnidarian.

_____

_____

SUMMARIZE IT

Student Edition pp. 521–523

Hydras have something called a nerve net. **Describe** what it is and how it works.

_____

_____

_____

Student Edition p. 523

Describe how a sponge obtains nutrients.

_____

_____

_____

_____

Section 28.3   Diversity Among the Lophotrochozoans, *pp. 524–532*

EXTENDING KNOWLEDGE

This section takes the AP Essential Knowledge you have learned further, and may provide illustrative examples useful for the AP Exam.

REVIEW IT

Student Edition p. 524

Sort the phyla of the Lophotrochozoans into the correct groups: molluscs, phoronids, tapeworms, brachiopoda, rotifers, flatworms, annelids, bryozoans.

Trochozoans	Lophophoroans

Name three types of flatworms.

_____

Section 28.3 Diversity Among the Lophotrochozoans (continued)

Student Edition p. 525

USE IT

The organisms of the lophotrochozoa are very diverse in their anatomy and life cycles. What are some characteristics they all share in common?

_____

_____

_____

SUMMARIZE IT

Describe how parasitic flatworms have evolved to live within their hosts.

_____

_____

_____

Section 28.4 Diversity of the Ecdysozoans, *pp. 532–538*

EXTENDING KNOWLEDGE

This section takes the AP Essential Knowledge you have learned further, and may provide illustrative examples useful for the AP Exam.

REVIEW IT

Using the vocabulary on the left, **fill in the blanks** below.

Vocabulary

arthropods
cuticle
Ecdysozoan
insects
metamorphosis
pseudocoelom
roundworms
tracheae

Roundworms and arthropods belong to the group _____. Ecdyosozoans have an outer covering called a _____ which is both a protective shell and for structural support. The phylum Nematoda contain _____, nonsegmented worms which have a body cavity called a _____. Many organisms including _____, crustaceans, and spiders are _____. Arthropods have distinct features including an exoskeleton, nervous system, segmentation, and respiratory organs, such as air tubes called _____. Some arthropods undergo a drastic change during their life cycle called _____.

USE IT

Student Edition p. 532

The organisms of the Ecdysozoans are very diverse in their anatomy and life cycles. What are some characteristics they all share in common?

_____

_____

Section 28.4 Diversity of the Ecdysozoans (continued)

Student Edition p. 532

SUMMARIZE IT

What are some features which have led to the evolutionary success of arthropods?

_____

_____

_____

_____

Section 28.5 Invertebrate Deuterostomes, *pp. 539—540*

> ### EXTENDING KNOWLEDGE
>
> This section takes the AP Essential Knowledge you have learned further, and may provide illustrative examples useful for the AP Exam.

REVIEW IT

Student Edition p. 539

List two ways sea starts and chordates are related.

_____

_____

USE IT

Student Edition pp. 539—540

What are tube feet, and how are they used?

_____

_____

_____

SUMMARIZE IT

A sea star is a member of the Phylum Echinodermata. Write a short paragraph describing its life cycle, morphology, and physiology.

_____

_____

_____

_____

_____

These questions were posed in the *Biology* chapter opener (page 513). Answer them using the knowledge you've gained from this chapter.

1. What features are common to all animals, regardless of their complexity, lifestyle, and habitat?

_____

_____

_____

2. Where do invertebrates fit into the evolutionary history of animals?

_____

_____

_____

_____

3. How do *HOX* genes orchestrate the development of the body plan of animals, while also providing evidence for a common ancestor?

_____

_____

_____

29 Vertebrate Evolution

As you work through your AP Focus Review Guide, keep this chapter's Big Ideas in mind:

FOLLOWING THE BIG IDEAS

 BIG IDEA 1 Evolutionary trends in vertebrate animals move toward endothermy, a four-chambered heart, air breathing, and internal fertilization and fetal development.

 BIG IDEA 2 Animal body systems help maintain homeostasis but are tailored to specific needs and environments.

 BIG IDEA 4 Organ systems in animals promote efficiency in the use of matter and energy.

Section 29.1   The Chordates, *pp. 545—546*

Essential Knowledge Covered:
1.B.2

EK 1.B.2
Student Edition pp. 545—546

REVIEW IT

What do cephalochordates and urochorates have in common?

_____

_____

EK 1.B.2
Student Edition p. 545

USE IT

Identify and describe the structures numbered in the diagram below.

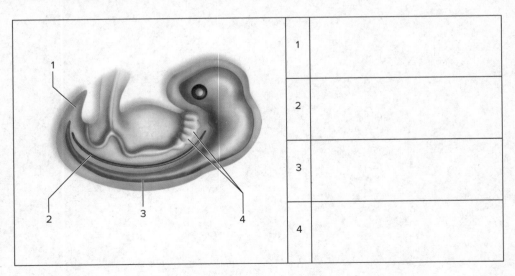

1	
2	
3	
4	

EK 1.B.2

SUMMARIZE IT

The structures in the diagram above are all seen at some point during development of all chordates. What does this indicate about all chordates?

_____

_____

Section 29.2   The Vertebrates, *pp. 547–548*

Essential Knowledge Covered:
1.A.4, 4.A.4

EK 1.A.4
Student Edition p. 547

REVIEW IT

How are vertebrates different from sea squirts or lancelets?

_____

_____

EK 1.A.4
Student Edition pp. 547–548

Define the following terms.

Tetrapod: _____

Gnathostome: _____

Amniote: _____

USE IT

EK 4.A.4
Student Edition p. 547

List three benefits of having an endoskeleton.

_____

_____

EK 1.A.4
Student Edition p. 547

Describe some traits found in fish fossils from the Ordovician which showed characteristics of adaptation to land.

_____

_____

SUMMARIZE IT

EK 1.A.4

Describe cephalization in vertebrates.

_____

_____

_____

Section 29.3   The Fishes, *pp. 548–550*

Essential Knowledge Covered:
2.B.2, 2.C.2, 2.D.2

REVIEW IT

Identify the fishes based on the description.

Description	Fishes
sharks, rays, skates: fishes with cartilaginous skeletons and lack gill covers	
the majority of living vertebrates	
extinct jawed fishes for the Devonian, thought to be the early ancestors of sharks and bony fishes	
bony fishes with fleshy fins supported by bones; includes lungfish	
bony fishes with fan-shaped fins	
agnathans	

USE IT

EK 2.C.2
Student Edition p. 548

How do most fishes regulate their temperature?

_____

_____

EK 2.D.2
Student Edition pp. 549–550

What is a swim bladder and how does it work?

_____

_____

_____

SUMMARIZE IT

EK 2.B.2
Student Edition p. 548

During field work, you come across a species of fish you've never seen before. How would you classify the fish as cartilaginous, bony, or lobe-finned?

_____

_____

_____

_____

Section 29.4 The Amphibians, *pp. 551—552*

Essential Knowledge Covered:
1.A.2, 4.C.2

EK 4.C.2
Student Edition p. 551

REVIEW IT

Identify the name of the characteristic that defines an amphibian.

Description	Characteristic
plays an active role in water balance, respiration, and thermal regulation	
eggs and sperm are deposited in the water	
sight, hearing, smell	
three chambered heart with a single ventricle and two atria	
depend on environment to regulate internal temperature	
developed pelvic and pectoral girdle	
respiratory organs which are supplemented with cutaneous respiration	

USE IT

EK 1.A.2
Student Edition pp. 551—552

Describe the three groups of amphibians.

_____

_____

_____

SUMMARIZE IT

EK 1.A.2

Amphibians have become very diverse. **Describe** an adaptation that has allowed a particular amphibian to thrive in its environment.

_____

_____

_____

_____

Section 29.5   The Reptiles, *pp. 553–558*

Essential Knowledge Covered:
1.B.2

EK 1.B.2
Student Edition p. 553

REVIEW IT

Describe three characteristics which have allowed reptiles to become fully adapted to land.

_____

_____

EK 1.B.2
Student Edition pp. 557–558

USE IT

Which traits are unique to birds compared to other reptiles, and which traits do they share?

_____

_____

_____

EK 1.B.2

SUMMARIZE IT

List three pieces of evidence which support that birds evolved from dinosaurs.

_____

_____

_____

Section 29.6   The Mammals, *pp. 559–561*

Essential Knowledge Covered:
1.B.1, 4.B.4

EK 1.B.1
Student Edition p. 559

REVIEW IT

List the three mammalian lineages.

_____

EK 1.B.1
Student Edition p. 560

Define placenta.

_____

_____

EK 1.B.1
Student Edition p. 559

USE IT

Describe the three characteristics that are unique to mammals.

_____

_____

Section 29.6   The Mammals (continued)

EK 1.B.1
Student Edition p. 560

Compare and contrast monotremes and marsupials.

_____

_____

_____

_____

EK 4.B.4

SUMMARIZE IT

Placental mammals are the most dominant group of mammals. What event allowed for this group to become so diverse?

_____

_____

_____

_____

AP Reviewing the Essential Questions

These questions were posed in the *Biology* chapter opener (page 544). Answer them using the knowledge you've gained from this chapter.

1. What characteristics distinguish vertebrates from other animals and facilitated their evolution from aquatic to terrestrial environments?

_____

_____

_____

2. How does vertebrate evolution relate to the Earth's changing environments?

_____

_____

_____

30 Human Evolution

As you work through your AP Focus Review Guide, keep this chapter's Big Ideas in mind:

FOLLOWING THE BIG IDEAS

 BIG IDEA 1 Fossil evidence suggests humans have evolved over millions of years, and that apes and humans shared a common ancestor.

Section 30.1 Evolution of Primates, *pp. 565–569*

Essential Knowledge Covered:
1.A.4, 1.B.1, 1.B.2

EK 1.B.1
Student Edition p. 565

REVIEW IT

Describe how primates are adapted for an arboreal life.

_____

_____

_____

EK 1.A.4
Student Edition p. 566

USE IT

Classify the non-prosimian primates.

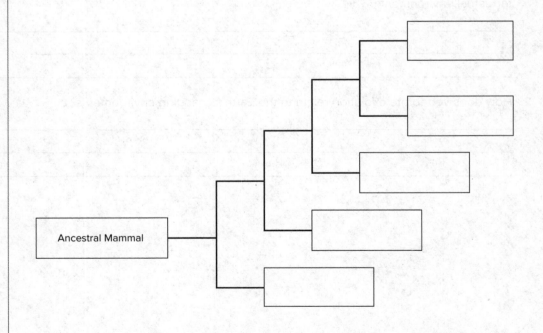

Ancestral Mammal

Section 30.1 Evolution of Primates (continued)

EK 1.A.4, 1.B.2

SUMMARIZE IT

Proconsul is a fossil from about 35 MYA. What does Proconsul represent and how does it compare to a current monkey skeleton?

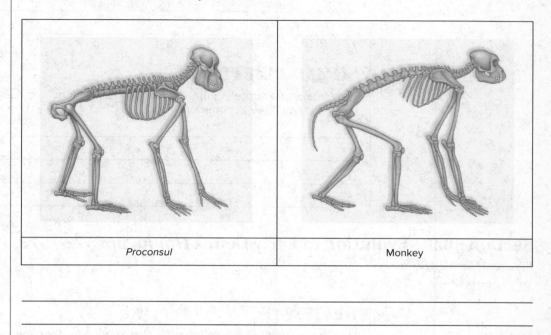

| Proconsul | Monkey |

_____

_____

_____

_____

Section 30.2 Evolution of Humanlike Hominins, *pp. 569–572*

EXTENDING KNOWLEDGE

This section takes the AP Essential Knowledge you have learned further, and may provide illustrative examples useful for the AP Exam.

REVIEW IT

Student Edition pp. 570–571

Identify the hominin.

Description	Hominin
• Slightly larger than a chimpanzee • Small head, brain- 370 to 515 cc • Fossils date back to 4 MYA • Walked erect, may have eaten meat • Used stone tools	
• About the size of a chimpanzee • Small head, brain- 300 to 350 cc • Fossils remnants date back to 5.6 MYA • Walked erect but spent time in trees • Had opposable big toe	

Section 30.2 Evolution of Humanlike Hominins (continued)

Student Edition p. 571

USE IT

Provide two examples of the advantages of bipedalism in early hominins.

_____

_____

SUMMARIZE IT

How are *Homo sapiens* different from Ardipithecines and Australopithecines? What traits do they have in common?

_____

_____

_____

Section 30.3 Evolution of Early Genus *Homo, pp. 572–573*

Essential Knowledge Covered:
1.B.2, 1.C.1, 1.C.2

REVIEW IT

EK 1.B.2
Student Edition pp. 572–573

Use the information given to you in this section to describe the traits of the early *Homo* species.

Homo floresiensis	
Homo ergaster	
Homo habilis and *Homo rudolfensis*	

SUMMARIZE IT

EK 1.C.1, 1.C.2

What might have led to the evolution of *Homo floresiensis*?

_____

_____

_____

Section 30.4   Evolution of Later Genus *Homo, pp. 573–577*

REVIEW IT

Using the vocabulary on the left, **fill in the blanks** below.

Vocabulary

biocultural evolution
Cro-Magnons
Denisovans
hunter-gatherers
Neandertals

biocultural evolution	
	Oldest fossils of *Homo sapiens*
Denisovans	
hunter-gatherers	
	archaic humans that lived 200,000–300,000 years ago

USE IT

Student Edition p. 575

Why did humans start agricultural societies?

_____

_____

_____

SUMMARIZE IT

Humans alive today show extreme variation in phenotypes. How did these variations arise?

_____

_____

_____

While populations of humans across the world do show variation in phenotypes, what evidence is there that we all arose from a common ancestor?

_____

_____

_____

_____

AP Reviewing the Essential Questions

These questions were posed in the *Biology* chapter opener (page 564). Answer them using the knowledge you've gained from this chapter.

1. How can an understanding of evolutionary patterns in other animals help us understand our evolutionary history?

   _____

   _____

   _____

2. What can the fossil record and comparative genomics tell us about human evolution?

   _____

   _____

   _____

   _____

31 Animal Organization and Homeostasis

As you work through your AP Focus Review Guide, keep this chapter's Big Ideas in mind:

FOLLOWING THE BIG IDEAS

 BIG IDEA 2 Homeostasis is essential for survival, with regulation and feedback mechanisms operating constantly to achieve it.

Section 31.1   Types of Tissues, *pp. 582–588*

Essential Knowledge Covered:
2.D.2, 2.D.4

EK 2.D.4
Student Edition p. 582

REVIEW IT

Define tissue.

_____

_____

EK 2.D.2
Student Edition pp. 586–587

Name the tissue type given its function.

Function	Tissue
binds and supports body parts	
receives stimuli and transmits nerve impulses	
covers body surfaces, lines body cavities and forms glands	
moves the body and its parts	

USE IT

EK 2.D.2
Student Edition p. 585

Describe two ways white blood cells fight off infection.

_____

_____

SUMMARIZE IT

EK 2.D.2

Read the feature "Regenerative Medicine" on page 588 of your textbook. How is regeneration related to homeostasis? How do organisms without regenerative ability maintain homeostasis?

_____

_____

_____

Section 31.2 Organs, Organ Systems, and Body Cavities, *pp. 589–590*

Essential Knowledge Covered:
2.B.2, 2.C.1, 2.D.1

EK 2.B.2
Student Edition p. 590

REVIEW IT

Define the following vocabulary words.

Body cavity – _____

_____

Organs – _____

_____

Organ system – _____

_____

EK 2.D.1
Student Edition p. 590

USE IT

The diagram below illustrates three human body cavities. **Identify** the cavity and list its contents.

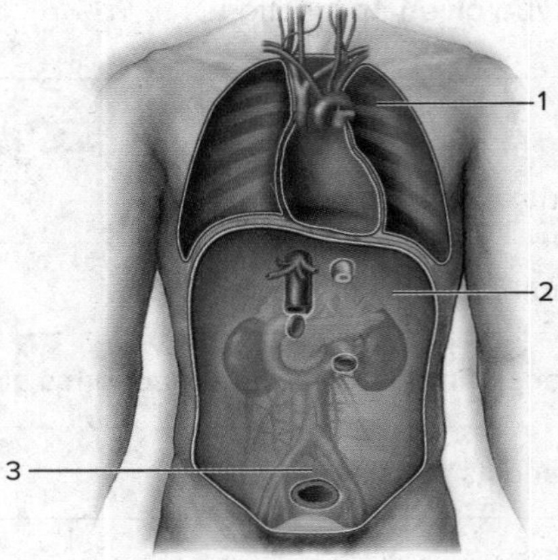

1.	
2.	
3.	

EK 2.C.1, 2.D.1

SUMMARIZE IT

List four organ systems and their corresponding life processes.

_____

_____

_____

_____

Section 31.3   The Integumentary System, *pp. 591—594*

Essential Knowledge Covered:
2.C.1, 2.C.2

Vocabulary

dermis
epidermis
hair follicles
melanocytes
nails
oil glands
skin
sweat glands

EK 2.C.2
Student Edition p. 592

EK 2.C.1
Student Edition p. 593

EK 2.C.2

REVIEW IT

Using the vocabulary on the left, **fill in the blanks** below.

The _____ covers the human body and has two main regions. The

_____ is made up of stratified squamous epithelium, and the _____

is a region of dense fibrous connective tissue. In the epidermis, specialized cells called

_____ produce a pigment called melanin; whereas _____, glands,

and _____ are largely part of the dermis. Humans have _____ and

_____ which secrete sebum and sweat respectively.

USE IT

How does UV radiation from the sun effect melanin production in the skin?

_____

_____

Provide a pro and a con of sun exposure to a light-skinned individual living in a
temperate zone.

_____

_____

SUMMARIZE IT

Describe how sweating cools your body.

_____

_____

_____

Section 31.4 Homeostasis, *pp. 594–596*

Essential Knowledge Covered:
2.C.1, 2.C.2, 2.D.2

EK 2.C.2, 2.D.2
Student Edition p. 594

REVIEW IT

Identify the type of homeostatic regulation.

Regulation	Definition
	Body temperature is regulated by a variety of internal mechanisms.
	Body temperature is regulated by the temperature of the external environment.

EK 2.D.2
Student Edition p. 594

USE IT

Given the organ system, **describe** a role it plays in homeostasis.

Organ	Homeostatic regulation
liver	
respiratory system	
kidneys	
digestive system	

EK 2.C.1, 2.C.2

SUMMARIZE IT

Using the example of the regulation of temperature in the human body, **describe** what happens when the body falls below normal temperature. What type of feedback mechanism is this?

_____

_____

_____

_____

AP Reviewing the Essential Questions

These questions were posed in the *Biology* chapter opener (page 581). Answer them using the knowledge you've gained from this chapter.

1. What is the difference between negative and positive feedback mechanisms in the regulation of homeostasis? What is an example of each?

   _____

   _____

   _____

   _____

2. How do specialized tissues, organs, and organ systems allow animals to better adapt to their environment?

   _____

   _____

   _____

32 Circulation and Cardiovascular Systems

As you work through your AP Focus Review Guide, keep this chapter's Big Ideas in mind:

FOLLOWING THE BIG IDEAS

BIG IDEA 1 Evolutionary trends in vertebrate animals move toward a closed circulatory system and a four-chambered heart.

BIG IDEA 2 Feedback mechanisms throughout the cardiovascular system allow extreme responses with an eventual return to homeostasis.

BIG IDEA 4 The circulatory system connects with every other organ system in the body.

Section 32.1   Transport in Invertebrates, *pp. 601—602*

Essential Knowledge Covered:
2.D.2, 4.B.2

EK 4.B.2
Student Edition pp. 601—602

REVIEW IT

What is the difference between an open and a closed circulatory system?

_____

_____

_____

EK 4.B.2
Student Edition p. 601

List two functions of a circulatory system.

_____

_____

EK 4.B.2
Student Edition p. 601

USE IT

How do invertebrates without circulatory systems process nutrients?

_____

_____

_____

EK 2.D.2, 4.B.2

SUMMARIZE IT

How does oxygen enter the cells of a grasshopper?

_____

_____

_____

_____

Section 32.2 Transport in Vertebrates, *pp. 603–604*

Essential Knowledge Covered:
1.C.3, 4.A.4, 4.B.2

Vocabulary

arteries
arterioles
capillaries
cardiovascular system
pulmonary circuit
systemic circuit
veins
venules

EK 4.A.4, 4.B.2
Student Edition p. 603

REVIEW IT

Given the definition on the left, **fill in** the correct vocabulary word(s) on the right.

Definition	Vocabulary Word
Drain blood from the capillaries and join to form a vein	
Small arteries whose diameter are regulated by the nervous and endocrine system	
The closed circulatory system of vertebrates	
The two circuits of the vertebrate cardiovascular system	
Carry blood away from the heart	
Return blood to the heart	
Blood vessels which exchange materials with interstitial fluid	

USE IT

List the three types of blood vessels in vertebrate circulatory systems and **draw** a simple diagram showing the directions in which they carry blood or interstitial fluids in the body.

_____

_____

EK 1.C.3, 4.A.4, 4.B.2

SUMMARIZE IT

Compare and contrast the circulatory pathway of a fish and a bird.

_____

_____

_____

_____

_____

EK 4.A.4, 4.B.2

Describe the movement of blood and oxygen through the circulatory pathway in the illustration below.

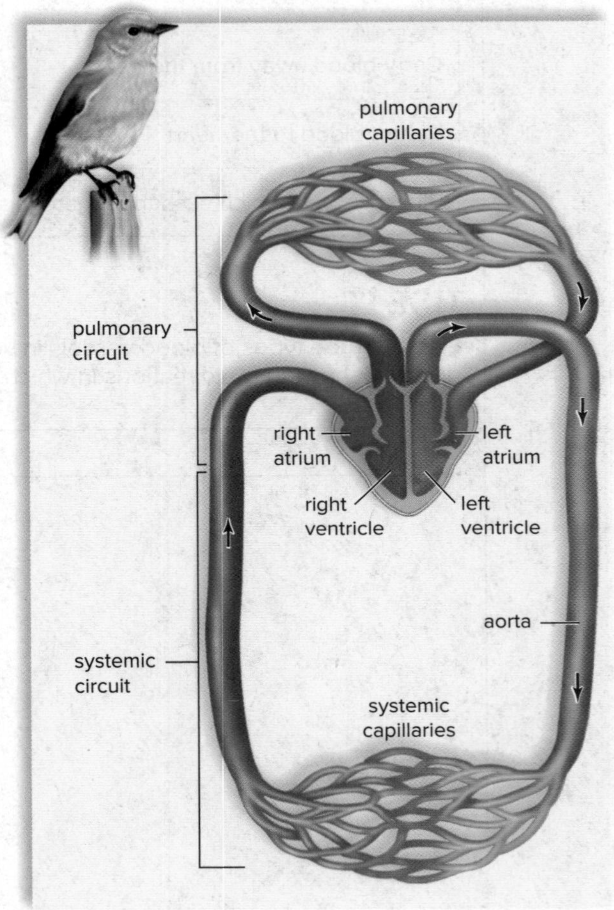

pulmonary capillaries

pulmonary circuit

right atrium

left atrium

right ventricle

left ventricle

aorta

systemic circuit

systemic capillaries

_____

_____

_____

_____

Section 32.3   The Human Cardiovascular System, *pp. 605–611*

Essential Knowledge Covered:
4.A.4, 4.B.2

EK 4.B.2
Student Edition p. 607

REVIEW IT

List four structures found in the human heart.

_____

_____

Fill in the missing phases of the cardiac cycle.

Time	Atria	Ventricles
0.15 sec		Diastole
0.30 sec	Diastole	
0.40 sec		Diastole

EK 4.B.2
Student Edition pp. 607–608

What causes the "lub-dub" sound of the heart beat?

_____

_____

_____

USE IT

EK 4.B.2
Student Edition p. 608

Identify the cardiovascular disease based on its description.

Description	Cardiovascular Disease
accumulation of soft masses of fatty materials between the inner linings of arteries which can trigger clots	
the blocking of an artery which causes a burning or squeezing sensation in the chest	
high blood pressure, narrowing of arteries	
the complete blocking of an artery in which part of the heart muscle dies due to lack of oxygen	

Section 32.3 The Human Cardiovascular System (continued)

SUMMARIZE IT

EK 4.B.2

Which node is called the pacemaker and why?

_____

_____

_____

EK 4.A.4

What internal systems can increase or decrease the rate and strength of heart contractions?

_____

_____

_____

Section 32.4   Blood, *pp. 613–618*

Essential Knowledge Covered:
2.C.1, 2.D.4, 4.A.4

REVIEW IT

Using the vocabulary on the left, **fill in the blanks** below.

Vocabulary

antibodies
antigen
formed elements
plasma
platelets
red blood cells
white blood cells

Blood consists of two main portions: _____ and the _____ such as cells and platelets. There are several plasma proteins including albumin which are the most plentiful, and _____ which are produced in the immune system. _____ transport oxygen and help transport carbon dioxide. Blood types are defined by the type of _____ found on red blood cells and in the plasma. _____ help fight infection, and _____ aid in clotting.

EK 4.A.4
Student Edition p. 614

List the five main types of white blood cells.

_____

_____

USE IT

EK 2.C.1
Student Edition p. 614

Compare and contrast the Rh and ABO system.

_____

_____

_____

_____

_____

Section 32.4   Blood (continued)

EK 2.C.1
Student Edition pp. 615–616

Describe how platelets are involved in blood clotting.

_____

_____

_____

_____

EK 2.D.4
Student Edition p. 615

What are T cell and B cells?

_____

_____

_____

EK 2.C.1
Student Edition p. 615

Draw a picture of how platelets and fibrin threads plug a punctured blood vessel.

EK 2.C.1, 2.D.4

SUMMARIZE IT

Describe how horseshoe crab blood clotting can identify a bacterial contamination.

_____

_____

_____

_____

These questions were posed in the *Biology* chapter opener (page 600). Answer them using the knowledge you've gained from this chapter.

1. How did the evolution of the four-chambered heart allow animals to adapt to challenges of terrestrial living?

   _____

   _____

   _____

2. How does the circulatory system help the cells of the body maintain homeostasis?

   _____

   _____

   _____

3. How does the circulatory system work with other body systems such as the respiratory or urinary?

   _____

   _____

   _____

33 The Lymphatic and Immune Systems

As you work through your AP Focus Review Guide, keep this chapter's Big Ideas in mind:

FOLLOWING THE BIG IDEAS

BIG IDEA 2	Rapid recognition and specialized destruction of foreign antigens are hallmarks of an advanced immune system.
BIG IDEA 3	Communication between immune cells ensures appropriate defensive action.
BIG IDEA 4	The extraordinary variety of cells that target and destroy antigens makes the immune system so successful.

Section 33.1   Evolution of Immune Systems, *pp. 622–623*

Essential Knowledge Covered:
2.D.4

EK 2.D.4
Student Edition p. 622

REVIEW IT

State the function of the immune system.

_____

_____

EK 2.D.4
Student Edition p. 622

Define *antigen*.

_____

_____

USE IT

EK 2.D.4
Student Edition p. 622

Identify two differences between innate immunity and adaptive immunity.

_____

_____

_____

SUMMARIZE IT

EK 2.D.4

Compare and contrast the immune system of a fruit fly and a human.

_____

_____

_____

_____

Section 33.2   The Lymphatic System, *pp. 623—625*

Essential Knowledge Covered:
2.D.4, 3.D.2

REVIEW IT

EK 2.D.4
Student Edition p. 623

Given the definition on the left, **fill in** the correct component of the lymphatic system on the right.

Definition	Component
a secondary lymphoid organ where lymph passes through and phagocytes engulf foreign debris and pathogens	
spongy, semisolid tissue where red blood cells are produced	
a secondary lymphoid organ where macrophages remove old and defective red blood cells	
tiny, closed-ended vessels found throughout the body which take-up excess interstitial fluid	
cells which mature in the thymus and fight infection	
the one-way system which drains fluid from tissue and returns it to the cardiovascular system	
a primary lymphoid organ where t cells mature	
interstitial fluid	
lymphocytes which mature in the red bone marrow	

USE IT

EK 2.D.4, 3.D.2
Student Edition p. 624

How do mature lymphocytes fight infections in the human body?

_____

_____

_____

EK 2.D.4
Student Edition p. 625

Why is a person without a spleen more susceptible to certain types of infections?

_____

_____

_____

Section 33.2   The Lymphatic System (continued)

EK 2.D.4

SUMMARIZE IT

Describe the three functions the lymphatic system.

_____

_____

_____

Section 33.3   Innate Immune Defenses, *pp. 625–628*

Essential Knowledge Covered:
2.D.3, 2.D.4, 3.D.2

REVIEW IT

Identify the component (or components) of the innate immune system.

Vocabulary

complement
dendritic cells
eosinophils
histamine
Inflammatory response
interferons
macrophages
mast cells
natural killer cells
neutrophils

Description	Component
phagocytes which leave the bloodstream and kill bacteria in tissue	
cytokines which affect the behavior of other cells	
large, granular lymphocytes that kill virus-infected or cancer cells on contact	
a series of events which occur after the infection of a pathogen	
two long-lived types of phagocytic white blood cells	
chemical mediators released by damaged cells which cause capillaries to dilate and become more permeable	
blood plasma proteins present in the blood plasma that are activated by pathogens and aid in immune response	
phagocytes which can also launch attacks against animal parasites such as tapeworms	
cells which release chemical mediators such as histamine	

USE IT

EK 2.D.3, 2.D.4
Student Edition pp. 626–627

Imagine you drop something heavy on your big toe. **Describe** the response that occurs in the tissue.

_____

_____

_____

EK 2.D.4
Student Edition p. 625

Identify the four types of innate defenses pictured below.

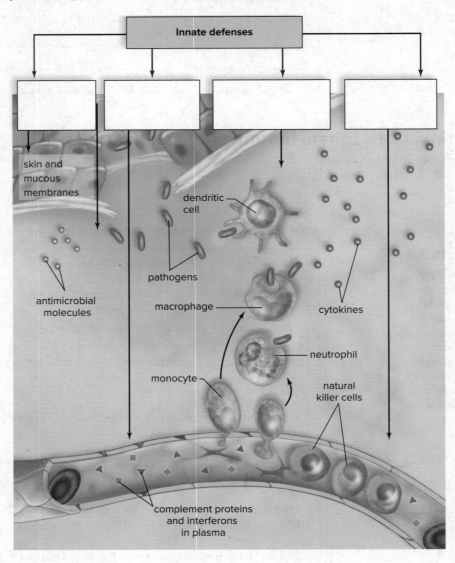

SUMMARIZE IT

EK 3.D.2

What might happen to a cell without a MHC-1 molecule on its surface?

_____

_____

_____

EK 2.D.4, 3.D.2

What are complement immune response proteins, and how can they help destroy pathogens?

_____

_____

_____

_____

Section 33.4 Adaptive Immune Defenses, *pp. 628–634*

Essential Knowledge Covered:
2.D.3, 2.D.4, 2.E.1, 3.B.2, 3.D.2, 3.D.4, 4.C.1

REVIEW IT

EK 2.D.4, 3.D.4, 4.C.1
Student Edition pp. 628–629

Given the vocabulary word on the right, give the definition.

Definition	Vocabulary Word
	immunoglobulins
	memory B cells
	helper T cells
	active immunity
	monoclonal antibodies
	Immunization

EK 2.D.4, 4.C.1
Student Edition p. 628

State the clonal selection theory.

_____

_____

USE IT

EK 3.D.2, 4.C.1
Student Edition p. 631

How are helper T cell and cytotoxic T cells different?

_____

_____

_____

EK 2.E.1, 3.B.2, 3.D.2, 4.C.1
Student Edition pp. 629–630

Draw and label a picture of a basic antibody. **Describe** how its antigen-binding sites plays a role in its function.

Section 33.4   Adaptive Immune Defenses (continued)

SUMMARIZE IT

EK 2.D.3, 2.D.4, 3.D.2

How does a T cell destroy a cell that has been infected with a virus?

_____

_____

_____

EK 2.D.3, 2.D.4, 2.E.1

Compare and contrast passive and active immunity.

_____

_____

_____

_____

_____

_____

Section 33.5   Immune System Disorders and Adverse Reactions, *pp. 636–638*

Essential Knowledge Covered:
2.D.3, 3.D.4, 4.C.1

REVIEW IT

Identify the immune disorder or adverse reaction.

Vocabulary

allergies
anaphylactic shock
asthma
autoimmune disease
delayed allergic response
immediate allergic response

hypersensitivity to substances such as pollen or food that would not ordinarily cause a reaction	
a sudden and life-threating drop in blood pressure due to an immediate allergic response	
a reaction initiated by memory t cells at the site of allergen contact	
an immediate reaction to a substance such as pollen or food caused by ige antibodies	
an inflammatory response that occurs in the lungs and causes wheezing	
when the immune system mistakenly attacks the body's own cells	

USE IT

EK 2.D.3, 3.D.4
Student Edition p. 636

Describe how an allergic reaction occurs.

_____

_____

SUMMARIZE IT

EK 2.D.3, 3.D.4

Choose an autoimmune disease described in this section and discuss how it may affect an individual.

_____

_____

_____

_____

_____

EK 4.C.1

How can MHC proteins cause a problem during organ transplants?

_____

_____

_____

_____

AP Reviewing the Essential Questions

These questions were posed in the *Biology* chapter opener (page 621). Answer them using the knowledge you've gained from this chapter.

1. What is the difference between innate immunity and adaptive immunity?

_____

_____

_____

2. What is a consequence to the organism if helper-T cells are destroyed by a pathogen, such as HIV?

_____

_____

_____

_____

3. How do both antibody-mediated immunity and cell-mediated immunity provide defense against exposure to foreign antigens? What types of cells are involved in each response?

_____

_____

_____

_____

34 Digestive Systems and Nutrition

As you work through your AP Focus Review Guide, keep this chapter's Big Ideas in mind:

FOLLOWING THE BIG IDEAS

 Increased surface area for digestion and absorption allows animals to obtain adequate nutrients.

 Compartmentalization and specialization of organs promote a more efficient disassembly line for digestion.

Section 34.1   Digestive Tracts, *pp. 642−644*

Essential Knowledge Covered:
2.A.3, 2.D.2

REVIEW IT

EK 2.A.3
Student Edition p. 642

List the four things a digestive system does.

_____

_____

EK 2.D.2
Student Edition p. 642

Describe a simple digestive tract.

_____

_____

USE IT

EK 2.D.2
Student Edition p. 642

Describe the difference between an incomplete digestive tract and a complete digestive tract.

_____

_____

_____

EK 2.D.2
Student Edition p. 644

What is a rumen, and how does it help cattle feed?

_____

_____

_____

EK 2.A.3, 2.D.2
Student Edition p. 642

Name an animal that has no digestive tract. How does it obtain nutrients?

_____

_____

_____

Section 34.1 Digestive Tracts (continued)

EK 2.D.2

SUMMARIZE IT

Describe two ways an organism may show adaptation to their diet.

_____

_____

_____

_____

Section 34.2   The Human Digestive System, *pp. 645–651*

Essential Knowledge Covered:
2.D.2, 4.A.4, 4.B.2

EK 4.A.4, 4.B.2
Student Edition pp. 645–648

REVIEW IT

Match the vocabulary word to its definition.

cirrhosis	A projection of the cecum which has an unclear function
polyps	Digested food and gastric juices delivered slowly to the small intestine
chyme	Small growths arising in the mucosa of the colon
mucosa	A chronic disease of the liver
bile	The central space of the digestive tract
lumen	A fluid made in the liver and stored in the gallbladder which emulsifies fat
appendix	The enzyme found in the salvia which digests starch
amylase	The innermost layer next to the lumen which produces mucus

EK 4.A.4
Student Edition p. 648

List three accessory organs of the human digestive system.

_____

_____

_____

USE IT

EK 2.D.2, 4.A.4
Student Edition pp. 647–648

Describe how the small intestine has an increased surface area for nutrient absorption.

_____

_____

EK 4.A.4, 4.B.2
Student Edition p. 648

What is the purpose of the large intestine?

_____

_____

EK 4.A.4, 4.B.2
Student Edition p. 649

How does the liver play a role in digestion?

_____

_____

_____

SUMMARIZE IT

EK 4.A.4, 4.B.2
Student Edition pp. 645–647

Explain how a piece of pizza is mechanically and chemically digested by a human.

_____

_____

_____

_____

_____

Section 34.3   Digestive Enzymes, *pp. 651–652*

Essential Knowledge Covered:
2.A.3, 4.A.4, 4.B.2

REVIEW IT

EK 4.A.4
Student Edition pp. 651–652

Identify the digestive enzyme.

breaks down proteins into peptides _____

aids in the conversion of maltose to glucose _____

the first enzyme to breakdown starch _____

enzymes from pancreatic juice that break down protein _____

breaks down peptides into amino acids _____

enzymes from pancreatic juice that break down starch _____

enzymes from pancreatic juice that break down fat _____

Section 34.3   Digestive Enzymes (continued)

EK 2.A.3
Student Edition p. 651

USE IT

What are the major nutritional components of food which digestive enzymes break down?

_____

_____

EK 4.B.2

SUMMARIZE IT

Describe what enzymes break down the piece of pizza you wrote about in Section 34.2.

_____

_____

_____

_____

_____

Section 34.4   Nutrition and Human Health, *pp. 652–656*

Essential Knowledge Covered:
2.A.3, 2.B.2, 2.C.2, 2.D.2

EK 2.A.3
Student Edition pp. 652–654

REVIEW IT

Identify the dietary component based on the description.

Description	Component
fats and oils which supply energy for cells	
twenty elements needed for various physiological functions, such as fluid balance	
eight molecules which adults cannot synthesizes but need for cellular processes	
organic compounds which regulate metabolic activities and are often part of coenzymes	
indigestible carbohydrates derived from plants	

EK 2.B.2, 2.C.2
Student Edition p. 654

Describe type 1 and type 2 diabetes.

_____

_____

_____

EK 2.A.3, 2.D.2
Student Edition p. 653

USE IT

Why is it suggested that people eat unsaturated fats in greater quantity than saturated fats?

_____

_____

_____

_____

EK 2.A.3

SUMMARIZE IT

Why is it important to maintain a healthy diet?

_____

_____

_____

_____

AP Reviewing the Essential Questions

These questions were posed in the *Biology* chapter opener (page 641). Answer them using the knowledge you've gained from this chapter.

1. What special features of the organs of the digestive system and interactions among these organs promote efficiency in the breakdown of food and absorption of nutrients?

   _____

   _____

   _____

   _____

2. How does the digestive system contribute to homeostasis?

   _____

   _____

   _____

   _____

35 Respiratory Systems

As you work through your AP Focus Review Guide, keep this chapter's Big Ideas in mind:

FOLLOWING THE BIG IDEAS

 BIG IDEA 2 The design of animal respiratory systems achieves maximum surface area for gas exchange.

 BIG IDEA 4 Cooperation between respiratory and circulatory systems is a mark of advanced animals.

Section 35.1   Gas-Exchange Surfaces, *pp. 660–664*

Essential Knowledge Covered:
2.A.3, 2.D.2, 4.A.4, 4.B.2

EK 2.A.3, 4.A.4
Student Edition pp. 660–664

REVIEW IT

Connect the vocabulary term to its definition.

respiration	organ that terrestrial vertebrates use to obtain oxygen
lungs	organ found in aquatic organisms for exchanging gases in water
tracheae	cavity connecting the oral and nasal cavities
gills	also known as the voice box
countercurrent exchange	gas exchange between a body's cells and the environment
pharynx	system of air tubs in insects
glottis	also known as the windpipe
larynx	tissue that covers the glottis when you swallow
trachea	process used by fish to transfer oxygen from the water into their blood
epiglottis	primary tubes that enter the lungs
bronchi	opening into the larynx in humans
bronchioles	air pockets within the lungs
alveoli	smaller passages within the lungs

Section 35.1   Gas-Exchange Surfaces (continued)

USE IT

EK 2.D.2
Student Edition p. 660

List three properties gas-exchange regions require in an organism in order for external respiration to be effective.

_____

_____

_____

SUMMARIZE IT

EK 2.A.3, 2.D.2, 4.B.2

Compare and contrast the bronchioles in humans to the tracheoles in insects.

_____

_____

_____

_____

EK 2.D.2

How have terrestrial vertebrates evolved to obtain oxygen?

_____

_____

_____

Section 35.2   Breathing and Transport of Gases, *pp. 665–669*

Essential Knowledge Covered:
2.D.2, 4.A.4, 4.B.2

REVIEW IT

EK 4.A.4
Student Edition pp. 665–668

Given the definition on the left, **fill in** the correct vocabulary word on the right.

Definition	Vocabulary Word
an iron-containing group found in hemoglobin	
the act of moving air out of the lungs	
a horizontal muscles that divides the thoracic cavity and the abdominal cavity	
the act of moving air into of the lungs	
the amount of pressure a gas exerts	

Section 35.2 Breathing and Transport of Gases (continued)

EK 4.A.4
Student Edition p. 665

USE IT

Identify which occurs during external respiration and which occurs during internal respiration.

Oxyhemoglobin gives up oxygen. _____

Oxygen binds with hemogoblin. _____

CO_2 enters red blood cells and becomes carbaminohemoglobin or a bicarbonate ion. _____

Carbonic acid is broken down with the aid of carbonic anhydrase into CO_2 and water. _____

SUMMARIZE IT

EK 4.A.4, 4.B.2

Draw a diagram illustrating the movement of carbon dioxide and oxygen through the respiratory and circulatory pathways in a human.

Section 35.3 Respiration and Human Health, *pp. 668–673*

Essential Knowledge Covered:
2.C.1, 2.D.3, 4.B.3

REVIEW IT

EK 2.D.3
Student Edition pp. 669–672

Diagnose the respiratory disease based on its description.

Description	Cardiovascular Disease
a genetic lung disease in which a faulty transport protein causes sticky mucus secretion that interfere with breathing	
a bacterial, viral, or fungal infection of the lungs which fill the bronchi or alveoli with pus or fluid	
a chronic and incurable lung disorder where alveoli are distended and damaged	
a tumor in the lungs	
the inflammation of the pharynx caused by a virus	

Section 35.3 Respiration and Human Health (continued)

EK 4.B.3
Student Edition p. 669

SUMMARIZE IT

Why is there no vaccine for the common cold?

_____

_____

_____

_____

AP Reviewing the Essential Questions

These questions were posed in the *Biology* chapter opener (page 659). Answer them using the knowledge you've gained from this chapter.

1. How do the respiratory and circulatory systems work together to supply all cells of the body with oxygen and eliminate carbon dioxide?

_____

_____

_____

_____

2. How does the respiratory system contribute to homeostasis? What are consequences if carbon dioxide levels are too high, or if oxygen levels are too low?

_____

_____

_____

_____

36 Body Fluid Regulation and Excretory Systems

As you work through your AP Focus Review Guide, keep this chapter's Big Ideas in mind:

FOLLOWING THE BIG IDEAS

 BIG IDEA 2   Differing environments will shape osmoregulation systems handed down from a common ancestor.

 BIG IDEA 4   Excretory organs must both extract wastes and maintain homeostasis by interactions inside and outside the system.

Section 36.1   Animal Excretory Systems, *pp. 678–681*

Essential Knowledge Covered:
2.D.2, 4.A.4, 4.B.2

EK 2.D.2
Student Edition p. 678

REVIEW IT

Define *osmoregulation*.

_____

_____

EK 2.D.2, 4.B.2
Student Edition pp. 679–680

Identify the nitrogenous waste excreted by the following organisms, and **draw** an arrow pointing from the compound that requires the most energy to synthesize to the compound that requires the least energy.

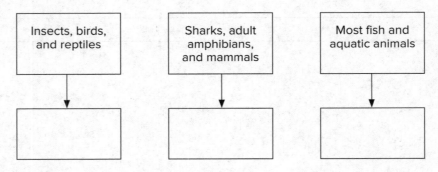

USE IT

EK 4.A.4
Student Edition p. 678

Why is excretion important to osmoregulation?

_____

_____

Section 36.1   Animal Excretory Systems (continued)

EK 4.A.4, 4.B.2
Student Edition p. 679

How do the kidneys help maintain homeostasis?

_____

_____

SUMMARIZE IT

EK 2.D.2, 4.B.2

How have marine mammals and sea birds adapted to living in a high salt environment?

_____

_____

_____

_____

_____

EK 2.D.2

Which drinks more water, a bony freshwater fish or a marine bony fish, and why?

_____

_____

_____

_____

_____

Section 36.2   The Human Urinary System, *pp. 681–688*

Essential Knowledge Covered:
2.C.1, 2.D.3, 4.A.4, 4.B.2

REVIEW IT

EK 4.A.4
Student Edition p. 681

Place in order the structures which urine travels through the human body: urethra, kidneys, ureter, urinary bladder.

_____

_____

EK 4.B.2
Student Edition pp. 682–685

List the three distinct processes of the urinary system.

_____

_____

Section 36.2 The Human Urinary System (continued)

USE IT

EK 4.A.4, 4.B.2
Student Edition pp. 684–685

Describe the three processes of the urinary system.

glomerular filtration	
tubular reabsorption	
tubular secretion	

EK 4.A.4, 4.B.2
Student Edition p. 685

List four reasons the kidneys are the primary organs of homeostasis in humans.

_____

_____

_____

SUMMARIZE IT

EK 2.C.1, 2.D.3

If a person becomes dehydrated, how does the body respond?

_____

_____

_____

EK 2.C.1, 4.A.4, 4.B.2

How do the liver, kidneys, and adrenal cortex work together to control the regulation of salt in the body?

_____

_____

_____

_____

These questions were posed in the *Biology* chapter opener (page 677). Answer them using the knowledge you've gained from this chapter.

1. Why is it important for all organisms, including animals, to maintain their normal water-salt balance?

   _____

   _____

   _____

2. How do the human excretory and circulatory systems work together to eliminate wastes while maintaining homeostatic levels of water and salt?

   _____

   _____

   _____

   _____

37 Neurons and Nervous Systems

As you work through your AP Focus Review Guide, keep this chapter's Big Ideas in mind:

FOLLOWING THE BIG IDEAS

 BIG IDEA 3   Neurons conduct electrochemical messaging throughout the animal body.

 BIG IDEA 4   The nervous system provides information input and output upon which the other systems rely for their operation.

Section 37.1   Evolution of the Nervous System, *pp. 692–694*

Essential Knowledge Covered:
3.E.2; 4.B.2

EK 4.B.2
Student Edition pp. 693–694

REVIEW IT

Given the definition on the left, **fill in** the correct vocabulary word on the right.

Definition	Vocabulary Word
a brain and spinal cord	
a simple nervous system organization in which neurons are in contact with one another and the cells of the body	
a concentration of nervous tissue in the anterior or head region	
all the nerves and ganglia that lie outside the central nervous system	
a cluster of neuron cell bodies	

USE IT

EK 3.E.2
Student Edition p. 692

List three functions of the nervous system.

_____

_____

_____

EK 4.B.2
Student Edition p. 693

How is the vertebrate brain organized?

_____

_____

_____

Section 37.1   Evolution of the Nervous System (continued)

EK 3.E.2
Student Edition pp. 692–693

List the organisms from least complex nervous system (1) to most complex (6) and describe the major characteristics of each.

Complexity	Organism	Nervous System
	Octopus	
	Cat	
	Hydra	
	Earthworm	
	Crab	

SUMMARIZE IT

EK 3.E.2, 4.B.2

What is the neocortex, what are its functions, and how is it different in humans than in other mammals?

_____

_____

_____

_____

Section 37.2   Nervous Tissue, *pp. 695–699*

Essential Knowledge Covered:
3.D.2, 3.D.3, 3.E.2, 4.B.2

REVIEW IT

EK 4.B.2
Student Edition pp. 695–696

List the three types of neurons:

_____

EK 4.B.2
Student Edition p. 695

Describe four neuroglia and their roles in the nervous system.

_____

_____

_____

_____

EK 3.D.3
Student Edition p. 696

USE IT

Describe what is happening in the neuron illustrated below.

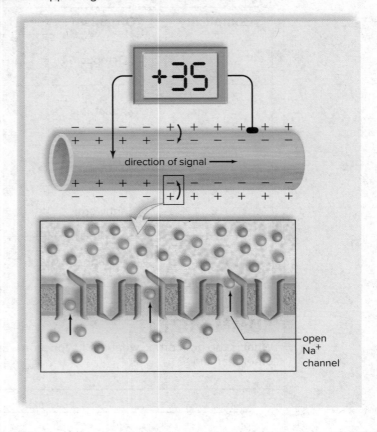

_____

_____

_____

_____

EK 3.D.2, 3.E.2
Student Edition p. 698

Describe the role the following neurotransmitters play in the body.

serotonin	
endorphins	
acetylcholine (ACh)	
norepinephrine	
dopamine	

Section 37.2   Nervous Tissue (continued)

EK 3.D.2, 4.B.2

SUMMARIZE IT

Using a diagram, **show** how a neurotransmitter transmits an action potential across a synapse.

Section 37.3   The Central Nervous System, *pp. 699–705*

Essential Knowledge Covered:
3.E.2, 4.B.2

EK 4.B.2
Student Edition pp. 702–705

REVIEW IT

Identify the structure of the Central Nervous System (CNS) given its function.

Structure	Function
	maintains homeostasis by regulating hunger, sleep, thirst, and body temperature
	the center of many reflex actions as well as means of communication between the brain and spinal nerves
	maintains posture and balance and coordinates muscle actions
	bundles of axons that bridge the cerebellum and the rest of the CNS
	accounts for sensation, voluntary movement, and all processes require for memory, learning, and speech
	regulates the heartbeat, breathing, and blood pressure
	the gatekeeper for sensory information en route to the cerebral cortex
	the two structures of the limbic system which are essential for learning and memory
	fills the spaces between the meninges to cushion and protect the CNS
	communicates and coordinates activities of the brain
	integrate motor commands

Section 37.3   The Central Nervous System (continued)

EK 3.E.2
Student Edition p. 699

USE IT

List the three specific functions of the Central Nervous System.

_____

_____

EK 3.E.2, 4.B.2

SUMMARIZE IT

Describe what parts of the brain allows you to recall something from a long time ago.

_____

_____

_____

EK 3.E.2, 4.B.2

Suppose you were walking in the woods and were startled by a snake slithering across the path. What part of the brain allowed you to be startled?

_____

_____

_____

Section 37.4   The Peripheral Nervous System, *pp. 706—709*

Essential Knowledge Covered:
3.D.1, 3.E.1, 3.E.2, 4.A.4, 4.B.2

EK 3.E.2
Student Edition pp. 706—709

REVIEW IT

List three types of nerves in the peripheral nervous system.

_____

_____

EK 3.E.1, 3.E.2, 4.A.4
Student Edition p. 708

USE IT

Describe why a person flinches when poked with a pin.

_____

_____

_____

_____

EK 3.D.1, 3.E.2, 4.A.4
Student Edition p. 709

What neurotransmitter is released in the sympathetic division of the PNS when someone is under attack and what does it do?

_____

_____

_____

EK 3.D.1, 3.E.2, 4.A.4
Student Edition p. 709

What type of reaction is illustrated in the diagram below?

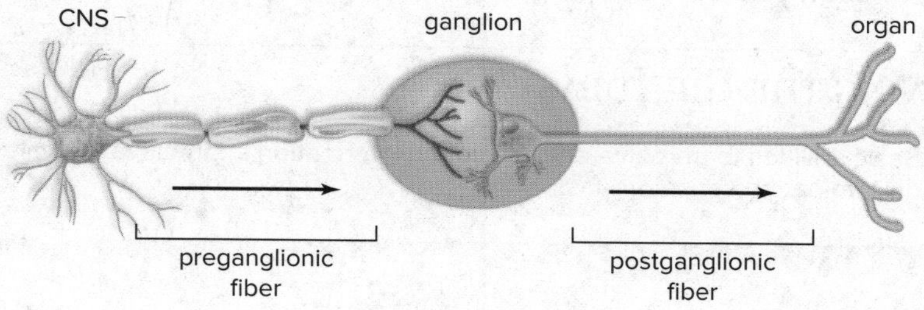

CNS   ganglion   organ

preganglionic fiber   postganglionic fiber

SUMMARIZE IT

EK 3.D.1, 3.E.1, 4.A.4

Compare and contrast the somatic motor and autonomic motor pathways.

_____

_____

_____

_____

_____

AP Reviewing the Essential Questions

These questions were posed in the *Biology* chapter opener (page 691). Answer them using the knowledge you've gained from this chapter.

1. What is the basic structure of the neuron, and how do changes in ion concentrations inside and outside of the neuron result in an action potential?

_____

_____

_____

2. How do neurotransmitters propagate nerve impulses across synapses?

_____

_____

_____

_____

38 Sense Organs

As you work through your AP Focus Review Guide, keep this chapter's Big Ideas in mind:

FOLLOWING THE BIG IDEAS

 BIG IDEA 1 Basic abilities to perceive environmental information in primitive animals have evolved into complex and discriminating sense organs.

Section 38.1   Sensory Receptors, *pp. 715–716*

Essential Knowledge Covered:
1.C.3

EK 1.C.3
Student Edition p. 715

EK 1.C.3
Student Edition p. 715

REVIEW IT

What is sensory transduction?

_____

Describe the sensory receptor and give an example.

Receptor	Function	Example
mechanoreceptors		
thermoreceptors		
electromagnetic receptors		
chemoreceptors		

EK 1.C.3

SUMMARIZE IT

Why do animals have sensory receptors?

_____

_____

_____

_____

Section 38.2   Chemical Senses, *pp. 716—718*

Essential Knowledge Covered:
1.B.1, 1.C.3

EK 1.B.1
Student Edition p. 716

REVIEW IT
Why is chemoreception thought to be the most primitive sense?

_____

_____

EK 1.C.3

SUMMARIZE IT
Compare and contrast the senses of taste and smell in humans.

_____

_____

_____

_____

_____

_____

Section 38.3   Sense of Vision, *pp. 718—724*

Essential Knowledge Covered:
1.B.1, 1.C.3

REVIEW IT
Using the vocabulary on the left, **fill in the blanks** below.

Vocabulary

camera-type eye
compound eyes
cone cells
conjunctiva
cornea
iris
lens
panoramic vision
photoreceptors
pupil
retina
rhodopsin
rod cells
stereoscopic vision

Sensory receptors that are sensitive to light are called _____. Arthropods

have _____; whereas, vertebrates and squid have _____. When

eyes face forward, organisms have _____ and _____.

In the human eye, there are three layers, the sclera which becomes the _____,

the _____ which keeps the eye moist, and the choroid which contains blood

vessels and pigment that absorbs stray light rays. The _____ regulates the size of the

pupil. The _____ regulates the light entering the eye, and the _____ helps form

images. Photoreceptors called _____ and _____ are found in the

_____ and allow us to see light and color. The pigment in the rods called

_____ plays a role in sending nerve impulse to the visual areas in the brain.

Section 38.3 Sense of Vision (continued)

EK 1.C.3
Student Edition p. 719

USE IT

Explain one reason is it beneficial for a rabbit to have panoramic vision.

_____

_____

EK 1.C.3
Student Edition p. 721

Why do cones provide us with a sharper image of an object than the rod cells?

_____

_____

_____

EK 1.B.1
Student Edition pp. 720–721

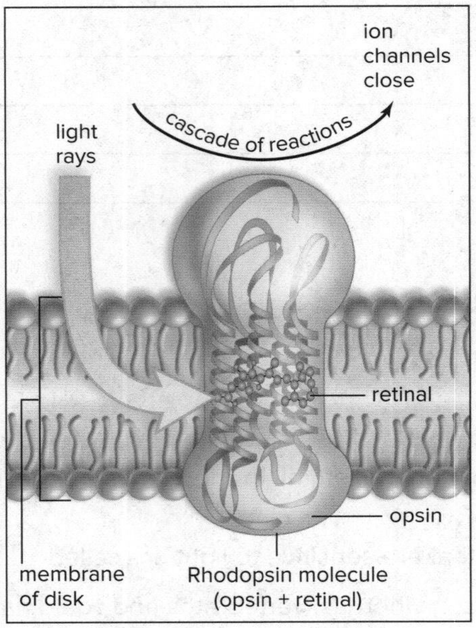

light rays

cascade of reactions

ion channels close

retinal

opsin

membrane of disk

Rhodopsin molecule (opsin + retinal)

Using the illustration on the left, **describe** how rhodopsin works.

_____

_____

_____

_____

_____

SUMMARIZE IT

EK 1.B.1, 1.C.3

Compare and contrast photoreceptors in planarians, dragonflies, and primates.

_____

_____

_____

_____

_____

_____

Section 38.4   Senses of Hearing and Balance, *pp. 724–728*

Essential Knowledge Covered:
1.B.1, 1.C.3

EK 1.B.1, 1.C.3
Student Edition pp. 724–726

REVIEW IT

Define the following vocabulary words.

middle ear	
gravitational equilibrium	
the organ of Corti	
inner ear	
rotation equilibrium	
outer ear	
auditory tube	
lateral line	

USE IT

EK 1.C.3
Student Edition p. 724

List the two sensory functions of the human ear.

_____

EK 1.C.3
Student Edition p. 727

How do fish stay upright in the water?

_____

_____

_____

SUMMARIZE IT

EK 1.B.1, 1.C.3

Compare and contrast what happens inside your ear when you shake your head 'no' versus nod your head 'yes'.

_____

_____

_____

_____

_____

_____

Section 38.5 Somatic Senses, *pp. 729–730*

SUMMARIZE IT

Fill in the charts to describe the somatic senses.

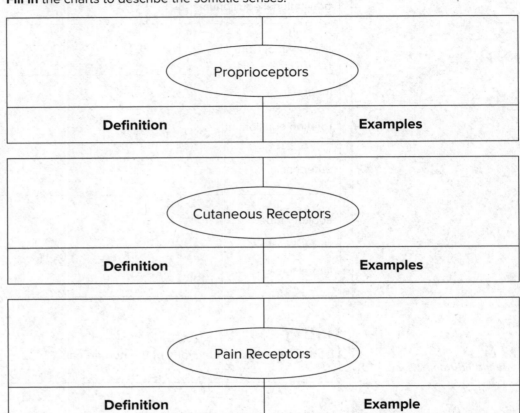

Proprioceptors	
Definition	**Examples**

Cutaneous Receptors	
Definition	**Examples**

Pain Receptors	
Definition	**Example**

AP Reviewing the Essential Questions

These questions were posed in the *Biology* chapter opener (page 714). Answer them using the knowledge you've gained from this chapter.

1. How do sensory organs and the nervous system work together to coordinate responses to stimuli?

_____

_____

_____

2. What is an example of a sensory receptor, and what features allow it to receive information from the environment?

_____

_____

_____

39 Locomotion and Support Systems

As you work through your AP Focus Review Guide, keep this chapter's Big Ideas in mind:

FOLLOWING THE BIG IDEAS

BIG IDEA 4 — The contraction of muscles is dependent upon their close relationship with the nervous system.

Section 39.1   Diversity of Skeletons, *pp. 734–735*

Essential Knowledge Covered:
4.A.4, 4.B.2

REVIEW IT

EK 4.B.2
Student Edition pp. 734–735

Provide a description for the three types of skeletal systems.

Skeleton	Description
hydrostatic skeleton	
endoskeleton	
exoskeleton	

USE IT

EK 4.B.2
Student Edition p. 734

What is the function of a skeletal system?

_____

_____

EK 4.A.4
Student Edition p. 735

List three advantages of a jointed endoskeleton.

_____

_____

_____

Section 39.1   Diversity of Skeletons (continued)

SUMMARIZE IT

EK 4.A.4

Describe four skeletal structures of mammals that are adapted to a particular mode of locomotion.

_____

_____

_____

_____

_____

_____

Section 39.2   The Human Skeletal System, *pp. 736–741*

Essential Knowledge Covered:
4.A.2, 4.A.4

REVIEW IT

EK 4.A.2, 4.A.4
Student Edition pp. 736–741

Match the vocabulary word to its definition.

joints	bone-forming cells
pectoral girdle	fibrous, cartilaginous, or synovial junctions between bones
pelvic girdle	upper portion of the appendicular skeleton, specialized for flexibility
osteoblast	bone-absorbing cells
red bone marrow	lower portion of the appendicular skeleton, specialized for strength
osteoclast	fibrous connective tissue that bind two bones together
osteocytes	tissue which produces red blood cells
ligaments	cells which affect the timing and location of bone remodeling

EK 4.A.4
Student Edition pp. 736–737

Name the two types of bone in a long bone.

_____

USE IT

EK 4.A.4
Student Edition p. 739

List two function of the vertebral column.

_____

_____

Section 39.2　The Human Skeletal System (continued)

SUMMARIZE IT

EK 4.A.4

Explain the roles of osteoblasts, osteoclasts, and osteocytes in bone growth and renewal.

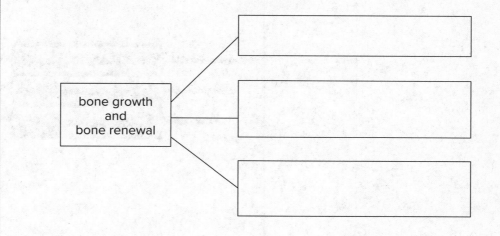

Describe three ways in which the skeletal system supports homeostasis.

_____

_____

_____

_____

Section 39.3　The Muscular System, *pp. 742–747*

Essential Knowledge Covered:
4.A.4, 4.B.2

REVIEW IT

Using the vocabulary words on the left, **fill in the blanks** below.

Vocabulary

actin
myofibrils
myosin
neuromuscular junction
oxygen debt
sarcolemma
sarcomeres
sarcoplasmic reticulum
sliding filament model
tendons
tetany
tone

Skeletal muscles are attached to the skeleton by _____. When a muscle is at rest, only some of the fibers in the muscle are contracting, and the muscle exhibits _____. Maximum sustained muscle contractions are known as _____. Muscle fibers are divided into contractile _____ which are made-up of _____. In the _____, thin or _____ filaments slide past thick _____ filaments, creating a muscle contraction. Calcium ions, stored in the _____, are essential the binding of myosin to actin. ATP supplies the energy for muscle contraction, and is generated through creatine phosphate breakdown and fermentation during _____. All muscle contractions are initiated when a nerve impulse to travel down from a motor neuron to a _____, causing acetylcholiune to bind to the _____.

EK 4.B.2
Student Edition p. 734

USE IT

Describe the state of the muscle in the illustration to the below. Identify the actin, myosin, and sarcomere.

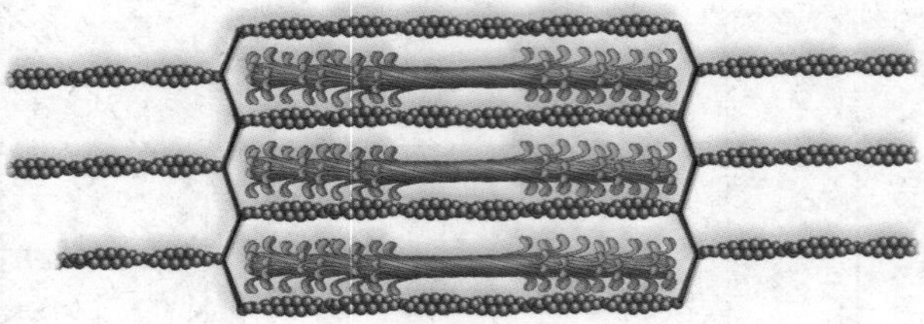

EK 4.A.4
Student Edition p. 747

What determines the duration of a muscle contraction?

_____

_____

SUMMARIZE IT

EK 4.A.4, 4.B.2

How does a nerve impulse translate into a muscle contraction? Be sure to identify the neurotransmitter, and the role of calcium and ATP in your answer.

_____

_____

_____

_____

_____

_____

These questions were posed in the *Biology* chapter opener (page 733). Answer them using the knowledge you've gained from this chapter.

1. What is the relationship between the ability of a muscle to contract and input from the nervous system?

   _____

   _____

   _____

   _____

2. How are the activities at the neuromuscular junction similar to the activities occurring at the synapse between neurons?

   _____

   _____

   _____

   _____

40 Hormones and Endocrine Systems

As you work through your AP Focus Review Guide, keep this chapter's Big Ideas in mind:

FOLLOWING THE BIG IDEAS

BIG IDEA 2   Endocrine imbalance leads to major physiological problems.

BIG IDEA 3   Hormones are chemical signals whose messages are received only by their target cells.

Section 40.1   Animal Hormones, *pp. 751–754*

Essential Knowledge Covered:
2.E.2, 3.D.1, 3.D.3

Vocabulary

chemical signals
first messenger
hormones
endocrine glands
endocrine system
pheromones
peptide hormones
second messenger
steroid hormones

REVIEW IT

Using the vocabulary on the left, **fill in the blanks** below.

The _____ regulates homeostasis through the production of chemicals from glands. The chemicals these glands secrete are known as _____. Unlike exocrine glands, _____ secrete their products into the blood stream and are distributed throughout the body. Hormones act as _____, passing messages between cells, body parts, and individuals. _____ are chemical signals that influence the behavior of others. Hormones derived from cholesterol are called _____, and hormones that are peptides, proteins, and other modified amino acids are _____. A hormone that never enters a cell but delivers the signal that a cascade of metabolic activity needs to occur is called the _____, and the molecule that receives this message and sets off the cascade is called the _____.

EK 3.D.1
Student Edition p. 753

What is cAMP? What does it do?

_____

_____

USE IT

EK 2.E.2, 3.D.3
Student Edition p. 752

List four major endocrine glands in the human body, the hormone they secrete, and the function the hormone plays.

_____

_____

Section 40.1   Animal Hormones (continued)

EK 2.E.2, 3.D.1
Student Edition p. 753

Describe two ways humans may be influenced by pheromones.

_____

_____

_____

_____

SUMMARIZE IT

EK 3.D.1

Compare and contrast how a peptide hormone and a steroid hormone activate a change in the body.

_____

_____

_____

_____

Section 40.2   Hypothalamus and Pituitary Gland, *pp. 755–757*

Essential Knowledge Covered:
2.C.1, 2.D.3, 2.E.2, 3.D.2, 3.E.2

REVIEW IT

EK 2.C.1
Student Edition p. 756

Draw a schematic (instructional sketch) of how the hypothalamus and pituitary gland work together.

Section 40.2   Hypothalamus and Pituitary Gland (continued)

EK 2.C.1, 2.E.2
Student Edition p. 755

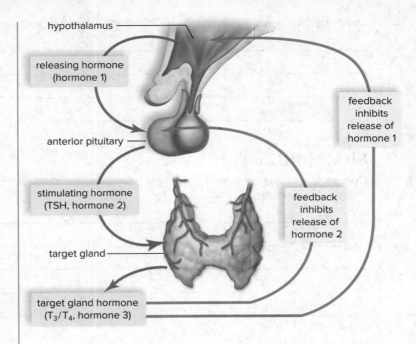

hypothalamus

releasing hormone
(hormone 1)

anterior pituitary

stimulating hormone
(TSH, hormone 2)

target gland

target gland hormone
(T₃/T₄, hormone 3)

feedback
inhibits
release of
hormone 1

feedback
inhibits
release of
hormone 2

USE IT

The figure to the left illustrates the pathway of how TSH is produced in the thyroid gland. **Explain** what is happening in the illustration.

_____

_____

_____

_____

SUMMARIZE IT

EK 2.C.1, 2.E.2, 3.D.2, 3.E.2

Explain the difference between a positive and negative feedback system in hormone production and give an example of each.

_____

_____

_____

_____

_____

EK 2.D.3

Describe the effects of the production of too little versus too much human growth hormone (GH) during childhood.

_____

_____

_____

_____

Section 40.3   Other Endocrine Glands and Hormones, *pp. 758–767*

Essential Knowledge Covered:
2.C.1, 2.D.2, 2.E.2, 3.D.1, 3.D.4

Vocabulary

adipose tissue
adrenal cortex
adrenal medulla
arachiodonate
cardiac cells
ovaries
pancreas
parathyroid
pineal gland
testes
thymus
thyroid

REVIEW IT

Describe the function of the hormone and identify the gland or organ it comes from.

Hormone	Function	Gland/Organ
aldosterone		
atrial natriuretic hormone		
calcitonin		
cortisol and glucocorticoids		
epinephrine and norepinephrine		
estrogen and progesterone		
glucagen		
insulin		
leptin		
melatonin		
parathyroid hormone		
prostagladin		
testosterone and androgens		
thyroxine		

Section 40.3 Other Endocrine Glands and Hormones (continued)

EK 2.C.1, 2.D.2, 3.D.4
Student Edition p.759

USE IT

Describe how calcitonin and PTH blood calcium levels.

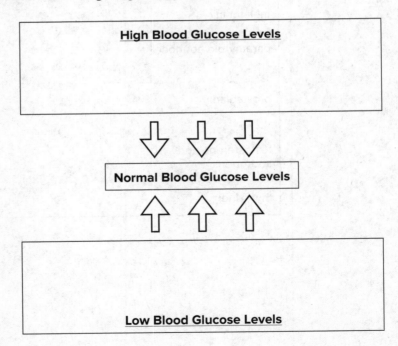

High Blood Calcium Levels

Normal Blood Calcium Levels

Low Blood Calcium Levels

EK 2.C.1, 2.D.2, 3.D.4
Student Edition p.763

Describe how insulin and glucagon regulate blood glucose levels.

High Blood Glucose Levels

Normal Blood Glucose Levels

Low Blood Glucose Levels

Section 40.3 Other Endocrine Glands and Hormones (continued)

SUMMARIZE IT

EK 2.C.2, 3.D.1, 3.D.4

List two ways the adrenal medulla responds to short term stress.

_____

_____

_____

EK 2.C.2, 3.D.1, 3.D.4

List two ways the adrenal medulla responds to long term stress.

_____

_____

_____

EK 2.C.2, 2.E.2, 3.D.1

a. experimental

b. winter

6 P.M.   6 A.M.

c. summer

Describe the level of melatonin production in relationship to the amount of light illustrated in each of the graphs on the left. Melatonin is depicted by the blue line, the yellow represents the amount of light, and the gray represents darkness.

_____

_____

_____

_____

_____

_____

AP Reviewing the Essential Questions

These questions were posed in the *Biology* chapter opener (page 750). Answer them using the knowledge you've gained from this chapter.

1. How does negative and positive feedback regulate hormone production and activity?

_____

_____

_____

2. How does the mode of action of a hormone differ from that of a neurotransmitter?

_____

_____

_____

41 Reproductive Systems

As you work through your AP Focus Review Guide, keep this chapter's Big Ideas in mind:

FOLLOWING THE BIG IDEAS

 BIG IDEA 2 Reproductive systems must respond to feedback from both internal and external signals.

Section 41.1 How Animals Reproduce, *pp. 771–772*

Essential Knowledge Covered:
2.C.1, 2.E.2

Vocabulary

copulation
germ cells
gonads
hermaphroditic
parthenogenesis
yolk

REVIEW IT

Identify the following vocabulary word from its definition.

Definition	Vocabulary Word
sexual union that facilitates the reception of sperm	
a nutrient-rich material inside an egg	
in which an unfertilized egg develops into a complete individual	
specialized organs which produce gametes	
having both male and female sex organs in a single body	
cells which give rise to eggs and sperm	

EK 2.E.2
Student Edition pp. 771–772

Provide a definition and an example of an organism for the following life history strategy.

Life History Strategy	Definition	Example Organism
oviparous		
ovoviviparous		
viviparous		

Section 41.1   How Animals Reproduce (continued)

EK 2.C.1
Student Edition p. 771

USE IT

While most animals are dioecious, some coral fish exhibit sequential hermaphroditism. **Describe** what happens with these coral reef fish.

_____

_____

SUMMARIZE IT

EK 2.E.2

Provide one advantage and one disadvantage for asexual reproduction in animals.

_____

_____

_____

_____

EK 2.C.1

How did shelled eggs allow for animals to colonize land?

_____

_____

_____

Section 41.2   Human Male Reproductive System, *pp. 773–776*

Essential Knowledge Covered:
2.C.1, 2.E.2

REVIEW IT

Fill in the function or the vocabulary word missing in the chart of the human male reproductive system.

Vocabulary

epididymis
prostate gland
semen
seminal vesicles
seminiferous tubules
sperm
testosterone
vas deferens

Component	Function
vas deferens	
	primary male sex organs
	conducts sperm and urine
prostate gland	
epididymis	
	provides large surface area for sperm development
	the main sex hormone in males
semen	
seminal vesicles	
	male gametes

Section 41.2   Human Male Reproductive System (continued)

EK 2.C.1
Student Edition p. 776

USE IT

List the three parts of a mature sperm and their corresponding functions.

_____

_____

_____

SUMMARIZE IT

EK 2.E.2

Draw a diagram showing how the hypothalamus is involved in the production of sperm.

Section 41.3   Human Female Reproductive System, *pp. 777–781*

Essential Knowledge Covered:
2.C.1

REVIEW IT

Fill in the function or the vocabulary word missing in the chart of the human female reproductive system.

Vocabulary

corpus luteum
endometrium
estrogen
follicle
oocyte
ovaries
progesterone
uterus
vagina

Component	Function
	the female sex hormones
oocyte	
	the birth canal
	houses developing embryo
follicle	
	the uterine lining where an embryo becomes inplanted
corpus luteum	
	primary female sex organs

Section 41.3   Human Female Reproductive System (continued)

EK 2.C.1
Student Edition p. 779

USE IT

Describe how the ovarian cycle is driven by hormone levels, and how this drives the uterine cycle.

Ovarian cycle

developing follicles   mature follicle   Ovulation   corpus luteum

Follicular Phase   **Luteal Phase**

_____

_____

_____

_____

SUMMARIZE IT

EK 2.C.1

Draw a diagram showing how the hypothalamus is involved in the production of oocytes.

Section 41.4   Control of Human Reproduction, *pp. 781–783*

Essential Knowledge Covered:
2.C.1

SUMMARIZE IT

EK 2.C.1

List four methods of birth control.

_____

_____

EK 2.C.1

Describe two reproductive technologies.

_____

_____

_____

_____

Section 41.5   Sexually Transmitted Diseases, *pp. 785−789*

Essential Knowledge Covered:
2.C.1

SUMMARIZE IT

EK 2.C.1

List three viral sexual transmitted diseases.

_____

_____

EK 2.C.1

List three bacterial sexual transmitted diseases.

_____

_____

EK 2.C.1

What is highly active antiretroviral therapy and why has it become effective in the fight against the spread of AIDS?

_____

_____

_____

_____

_____

_____

AP Reviewing the Essential Questions

These questions were posed in the *Biology* chapter opener (page 770). Answer them using the knowledge you've gained from this chapter.

1. What is the relationship between the hypothalamus/pituitary and the male and female reproductive systems?

_____

_____

_____

2. How do feedback mechanisms regulate gamete production?

_____

_____

42 Animal Development and Aging

As you work through your AP Focus Review Guide, keep this chapter's Big Ideas in mind:

FOLLOWING THE BIG IDEAS

 Control and timing of developmental events are achieved by differential gene expression.

 Transmission of various signals control gene expression and cell differentiation.

Section 42.1   Early Developmental Stages, *pp. 794–797*

Essential Knowledge Covered:
2.E.1, 3.D.2

Vocabulary

blastocoel
blastopore
blastula
cleavage
development
ectoderm
endoderm
embryo
fertilization
gastrula
gastrulation
germ layers
mesoderm
morula
neural plate
neural tube
notochord

EK 2.E.1, 3.D.2
Student Edition pp. 794–795

REVIEW IT

Using the vocabulary on the left, **fill in the blanks** below.

The formation of a diploid zygote forms as a result of _____. All changes that occur during the life cycle of an organisms is known as _____. During cellular development, the _____ undergoes cell divisions without growth, known as _____. This process forms a solid _____, and then becomes a hollow _____ with a fluid-filled cavity known as a(n) _____. Next, during tissue stages of development, _____ occurs forming a(n) _____ from the blastula. The early gastrula has two layers of cells, (1) the _____ and (2) the _____. The middle layer that forms is called the _____. Collectively, these three layers are called the _____. The _____ becomes the anus. Organ development occurs last. When the nervous system forms, the _____ emerges from the mesoderm. Above this, cells thicken to form a(n) _____ which becomes the _____, giving rise to the brain.

USE IT

Describe the events that unfold when a sperm makes its way through the corona radiata of the oocyte.

_____

_____

_____

_____

Section 42.1　Early Developmental Stages (continued)

EK 2.E.1
Student Edition p. 796

Identify the germ layer in which the following structures develop from.

the nervous system　　　　　　　　　＿＿＿＿＿＿＿＿＿

the thyroid and parathyroid glands　　＿＿＿＿＿＿＿＿＿

tooth enamel　　　　　　　　　　　　＿＿＿＿＿＿＿＿＿

lining of the digestive tract　　　　　＿＿＿＿＿＿＿＿＿

the cardiovascular system　　　　　　＿＿＿＿＿＿＿＿＿

hair and nails　　　　　　　　　　　　＿＿＿＿＿＿＿＿＿

the outer layer of the digest system　＿＿＿＿＿＿＿＿＿

EK 2.E.1
Student Edition p. 795

Draw a diagram of how a zygote develops into a morula, a blastula, and then forms three germ layers.

SUMMARIZE IT

EK 2.E.1

Compare and contrast the tissue stage of development between a frog and lancelet.

＿＿＿＿＿＿＿＿＿＿＿＿＿＿＿＿＿＿＿＿＿＿＿＿＿＿＿＿＿＿＿＿＿＿＿＿＿＿＿

＿＿＿＿＿＿＿＿＿＿＿＿＿＿＿＿＿＿＿＿＿＿＿＿＿＿＿＿＿＿＿＿＿＿＿＿＿＿＿

＿＿＿＿＿＿＿＿＿＿＿＿＿＿＿＿＿＿＿＿＿＿＿＿＿＿＿＿＿＿＿＿＿＿＿＿＿＿＿

＿＿＿＿＿＿＿＿＿＿＿＿＿＿＿＿＿＿＿＿＿＿＿＿＿＿＿＿＿＿＿＿＿＿＿＿＿＿＿

＿＿＿＿＿＿＿＿＿＿＿＿＿＿＿＿＿＿＿＿＿＿＿＿＿＿＿＿＿＿＿＿＿＿＿＿＿＿＿

＿＿＿＿＿＿＿＿＿＿＿＿＿＿＿＿＿＿＿＿＿＿＿＿＿＿＿＿＿＿＿＿＿＿＿＿＿＿＿

Section 42.2 Developmental Processes, *pp. 798–802*

Essential Knowledge Covered:
2.E.1, 3.B.2, 3.D.2

Vocabulary

apoptosis
cellular differentiation
cytoplasmic segregation
homeobox
homeodomain
homeotic genes
induction
maternal determinants
pattern formation
totipotent

REVIEW IT

Identify the vocabulary word.

the ability of a zygote to generate the entire organism	
a functionally important 60 amino-acid sequence in a homeobox	
the parceling out of maternal determinants during mitosis	
how tissues and organs are arranged in the body	
the ability for one embryonic tissue to influence the development of another	
a structural feature in a homeotic gene that all organisms share	
when cells become specialized in structure and function	
programmed cell death	
substances in the cytoplasm which influence the course of development	
a gene which selects for segmental identity	

USE IT

EK 2.E.1, 3.B.2
Student Edition p. 798

Scientists classify the cytoplasm of a frog egg as polar. What does this mean, and why is this important to development?

_____

_____

_____

_____

EK 2.E.1, 3.B.2. 3.D.2
Student Edition pp. 798–799

Compare and contrast maternal determinants and induction.

_____

_____

_____

_____

Section 42.2 Developmental Processes (continued)

EK 2.E.1, 3.B.2
Student Edition p. 801

Provide an example of how apoptosis is important during development.

_____

_____

SUMMARIZE IT

EK 2.E.1, 3.B.2, 3.D.2

Answer the following questions concerning the diagram below.

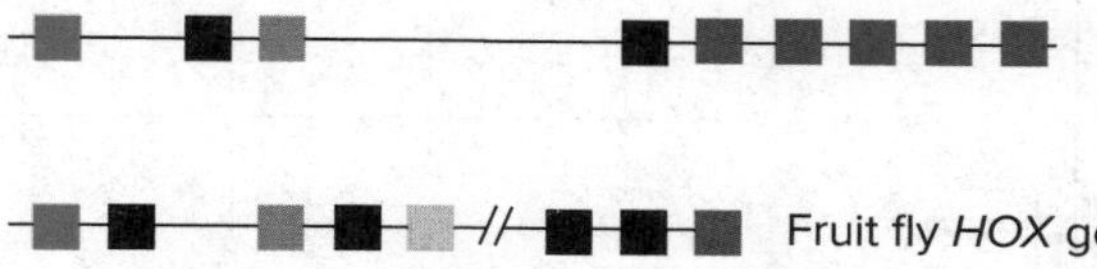

Mouse *HOX* genes

Fruit fly *HOX* genes

HOX genes control what portion of development?

_____

_____

Which genes occur in the same order in the diagram above?

_____

_____

What is the importance of the homeodomain in the homeotic genes?

_____

_____

_____

Section 42.3 Human Embryonic and Fetal Development, *pp. 802–808*

Essential Knowledge Covered:
2.E.1

REVIEW IT

Using the vocabulary on the left, **fill in the blanks** below.

Vocabulary

allantois
amnion
chorion
embryonic development
extraembryonic membranes
yolk sac

Human development can be divided into _____ and fetal development. _____ in humans include the chorion, the amnion, the allantois, and the yolk sac. The _____ provides nourishment, the _____ collects nitrogenous waste, the _____ protects the developing embryo, and the _____ carries out gas exchange.

Section 42.3   Human Embryonic and Fetal Development (continued)

EK 2.E.1
Student Edition pp. 804–807

USE IT

Describe what occurs during the weeks of embryonic development. Use the words: *umbilical cord, HCG, blastocyst, placenta, implantation, trophoblast*

Week	Development
one	
two	
three	
four and five	
six-eight	

SUMMARIZE IT

EK 2.E.1

What is the importance of the placenta?

_____

_____

_____

EK 2.E.1

List the three stages of birth.

_____

_____

Section 42.4   The Aging Process, *pp. 809–812*

EXTENDING KNOWLEDGE

This section takes the AP Essential Knowledge you have learned further, and may provide illustrative examples useful for the AP Exam.

Student Edition p. 809

REVIEW IT

Describe what would happen if animals did not age or die.

_____

_____

Section 42.4   The Aging Process (continued)

Student Edition pp. 809–811

USE IT

Identify an example of aging on the following organ systems.

sensory system	
integumentary system	
musculoskeletal system	
immune system	
reproductive system	
cardiovascular system	
nervous system	

SUMMARIZE IT

Identify the two major hypotheses concerning why we age.

_____

_____

AP Reviewing the Essential Questions

These questions were posed in the *Biology* chapter opener (page 793). Answer them using the knowledge you've gained from this chapter.

1. What is the relationship between targeted gene expression and the specialization of cells, tissues, and organs?

_____

_____

_____

2. What is the role of *HOX* and homeotic genes in pattern formation and overall development?

_____

_____

_____

3. How do maternal determinants and induction contribute to cellular differentiation?

_____

_____

_____

43 Behavioral Ecology

As you work through your AP Focus Review Guide, keep this chapter's Big Ideas in mind:

FOLLOWING THE BIG IDEAS

 Both innate and learned behaviors allow all organisms to adjust to individual and environmental changes.

 External communication of all types allows organisms to function successfully in their community.

Section 43.1   Inheritance Influences Behavior, *pp. 818–820*

Essential Knowledge Covered:
2.E.3, 3.E.1, 3.E.2

EK 3.E.1
Student Edition p. 818

REVIEW IT

Define *behavior.*

_____

_____

EK 2.E.3, 3.E.2
Student Edition p. 818

Explain the phrase "nurture versus nature."

_____

_____

EK 2.E.3
Student Edition p. 820

USE IT

How do marine snails demonstrate that behavior has genetic basis?

_____

_____

_____

SUMMARIZE IT

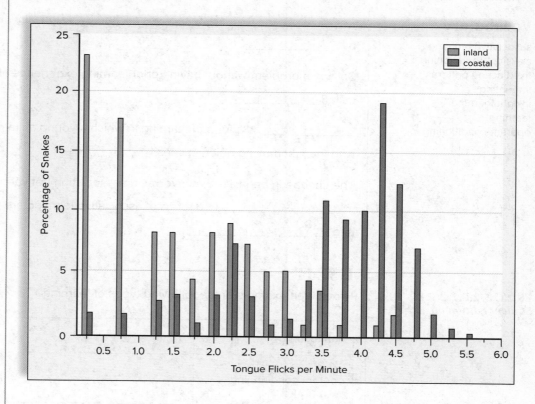

EK 2.E.3

This graph demonstrates the number of tongue flicks performed by two populations of garter snakes when exposed to slugs. Snakes use tongue flicks to "smell" their prey. **Provide** a reason costal snakes may eat slugs and inland snakes do not.

_____

_____

_____

_____

If the behavior of eating slugs in these garter snakes are indeed the result of a genetic basis, what would you expect the offspring of a coastal and inland snake to eat?

_____

_____

Section 43.2 The Environment Influences Behavior, *pp. 820−824*

Essential Knowledge Covered:
2.C.2, 2.E.1, 2.E.2, 3.E.1

REVIEW IT

Use the words *orientation*, *navigation*, and *migration* in a cohesive sentence.

_____

_____

EK 3.E.1
Student Edition pp. 820–824

Vocabulary

associative learning
classical conditioning
fixed action patterns
imprinting
insight learning
learning
operant conditioning

Fill in the blanks to complete the sentences below.

_____ are responses elicited by a sign stimulus but can be modified

by _____.

Solving a problem without having prior learning experience about the situation is

called _____.

_____ is a form of learning in which a young animal develops an association

with the first moving object it sees.

The change in behavior that involves an association between two event is known as

_____. Two forms of associative learning are _____

and _____.

USE IT

EK 2.C.2, 2.E.2, 2.E.1
Student Edition pp. 821–824

Provide an example of the following types of learning.

habituation	
operant conditioning	
insight learning	
classical conditioning	
imprinting	
associative learning	

EK 3.E.1
Student Edition p. 822

How do social interactions assist white-crowned sparrows learning how to sing?

_____

_____

_____

_____

Section 43.2   The Environment Influences Behavior (continued)

SUMMARIZE IT

EK 2.C.2, 2.E.2, 3.E.1

Using the flow diagram below, **describe** in general terms Pavlov's classical conditioning experiment.

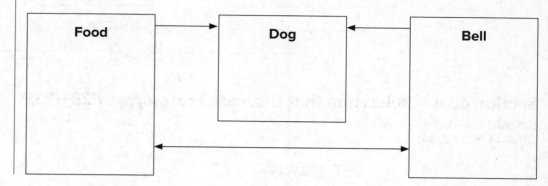

Section 43.3   Animal Communication, *pp. 824–828*

Essential Knowledge Covered:
2.C.2, 2.E.2, 3.D.1, 3.E.1

REVIEW IT

EK 2.C.2
Student Edition pp. 825–827

Determine the following types of communication.

Communication	Example
	a baboon opens its mouth really wide to warn another baboon to leave
	an ant marks its trail with a pheromone
	a bee performs a waggle dance to lead itself to food
	a bird calls an alarm to warn another bird

EK 2.C.2, 3.E.1
Student Edition p. 824

Define *society.*

_____

_____

USE IT

EK 3.E.1
Student Edition p. 824

List four reasons organisms communicate.

_____

_____

EK 2.E.2, 3.D.1
Student Edition p. 824

Describe two advantages of chemical communication.

_____

_____

Section 43.3   Animal Communication (continued)

EK 3.E.1

SUMMARIZE IT

Provide an example of visual communication.

_____

_____

_____

Section 43.4   Behaviors that Increase Fitness, *pp. 828–832*

Essential Knowledge Covered:
2.C.2, 2.E.2, 2.E.3, 3.E.1

REVIEW IT

Fill out the missing the vocabulary word or definition on the chart below.

Vocabulary

altruism
inclusive fitness
kin selection
monogamous
polyandrous
polygamous
reciprocal altruism

Vocabulary Word	Definition
altruism	
	an animal helps another animal but gains no immediate benefit but is repaid at some later time
inlusive fitness	
	the adptation to the environment due to reproductive success of the individuals relatives
polyandrous	
	the study of how natural selection shapes behavior
	a pair bonds to produce an offspring and both care for the young
polygamous	

EK 2.E.3
Student Edition p. 829

Describe the optimal foraging model.

_____

_____

USE IT

EK 2.E.3
Student Edition p. 829

Why would a shore crab prefer a medium sized mussel to eat instead of a large one?

_____

_____

SUMMARIZE IT

EK 2.C.2, 2.E.2, 3.E.1

The male bowerbird has developed a flamboyant display in order to make the female notice him. If the male becomes too intense in his display, however, what happens?

_____

_____

EK 2.E.3

Describe two ways in which altruism is beneficial.

_____

_____

_____

AP Reviewing the Essential Questions

These questions were posed in the *Biology* chapter opener (page 817). Answer them using the knowledge you've gained from this chapter.

1. How do genetics and the environment work together to influence both innate and learned behaviors?

_____

_____

_____

2. What are examples of strategies animals use to communicate information, and how do these strategies increase survival and reproductive success, i.e., fitness?

_____

_____

_____

3. What are advantages of cooperative behavior for both the individual and the community? Disadvantages?

_____

_____

_____

44 Population Ecology

As you work through your AP Focus Review Guide, keep this chapter's Big Ideas in mind:

FOLLOWING THE BIG IDEAS

BIG IDEA 2 Populations are impacted by both biotic and abiotic factors.

BIG IDEA 4 Though populations could grow exponentially, various limiting factors typically keep size at a sustainable level.

Section 44.1   Scope of Ecology, *p. 837*

Essential Knowledge Covered:
2.D.1

EK 2.D.1
Student Edition p. 837

SUMMARIZE IT

Place the following ecological levels in order from least complex to most complex: *community, biosphere, population, habitat, ecosystem.* **Provide** a definition and example for each.

Definition:	_____	Example:

Definition:	_____	Example:

Definition:	_____	Example:

Definition:	_____	Example:

Definition:	_____	Example:

Section 44.2 Demographics of Populations, *pp. 838–841*

Essential Knowledge Covered:
2.D.1, 4.A.5, 4.E.5

Vocabulary

age structure diagram
biotic potential
demography
limiting factors
population density
population distribution
rate of natural increase
resources
survivorship

REVIEW IT

Using the vocabulary on the left, **fill in the blanks** below.

The statistical study of a population is called _____. Studies can be performed on the number of individuals per unit area or _____, the pattern of dispersal of individuals or _____, or how the population is growing through the _____. Factors that are used to determine population distribution include _____ and _____, such as environmental aspects that determine where organisms live. The highest possible rate of natural increase for a population is known as its _____. The probability of how long an individual will live to a certain age is known as _____. The numbers of individuals alive in each generation of a population can mapped in a(n) _____.

USE IT

EK 2.D.1
Student Edition p. 838

List four important resources that humans need to survive.

_____

_____

EK 4.A.5
Student Edition p. 839

Calculate the rate of natural increase for a population of 2000 that has 50 births per year and 15 deaths. Recall that the rate of natural increase (*r*) is determined by the number of births per year minus the number of deaths per year divided by the number of individuals in a population.

EK 2.D.1, 4.E.5
Student Edition p. 839

Identify two limiting factors which may reduce a population's potential reproduction.

_____

_____

_____

SUMMARIZE IT

EK 2.D.1

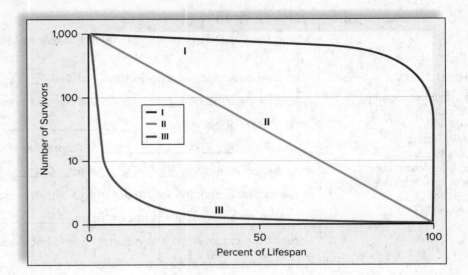

Identify if the statement pertains to survivorship curve I, II, or III shown on the graph above.

Most large mammals follow this type of survivorship curve. ___

Death is unrelated to age. ___

Death does not come until near the end of the lifespan. ___

Most individuals die very young. ___

Section 44.3   Population Growth Models, *pp. 841–844*

Essential Knowledge Covered:
2.D.1, 4.A.5, 4.B.3

EK 2.D.1, 4.A.5
Student Edition pp. 841–844

REVIEW IT

Define the models, curves, and factors of population growth.

exponential growth	
semiparity	
logistic growth	
iteroparity	
carrying capacity (*K*)	

Section 44.3   Population Growth Models (continued)

EK 2.D.1, 4.A.5
Student Edition p. 842

USE IT

Plot the following data on the line graph.

Generation	Population
1	4
2	4
3	10
4	20
5	40

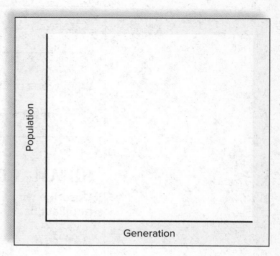

What is this type of growth curve called?

_____

What are the two phases of this type of growth curve?

_____

What types of factors are necessary for this type to growth to continue?

_____

_____

_____

EK 2.D.1, 4.A.5, 4.B.3
Student Edition p. 842

The graph below shows the number of yeast cells per hour growing in a beaker.

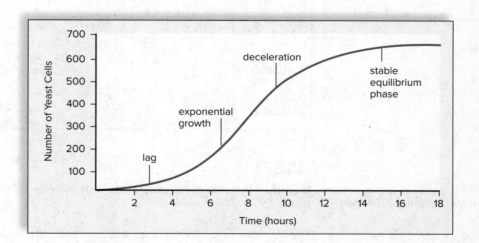

What is this type of growth called?

_____

Section 44.3   Population Growth Models (continued)

Describe what happens to yeast population during the four phases of its growth.

_____

_____

_____

What types of factors are necessary for this type to growth to continue?

_____

_____

SUMMARIZE IT

EK 2.D.1

Describe why a fishing company would want to monitor the carrying capacity of its fish populations.

_____

_____

_____

Section 44.4   Regulation of Population Size, *pp. 844–846*

Essential Knowledge Covered:
2.D.1, 2.D.3, 4.A.5, 4.B.3

REVIEW IT

EK 2.D.1, 2.D.3, 4.A.5
Student Edition pp. 844–846

Provide an example of the following.

density-independent factors	
density-dependent factors	
competition	
predation	

USE IT

EK 2.D.1, 2.D.3, 4.A.5
Student Edition pp. 844–845

Identify if the statement pertains to a density-dependent (D) or density-independent (I) factor.

The intensity of the effect does not increase with the increased population density. _____

It is usually a biotic factor _____

It is usually an abiotic factor _____

The intensity of the effect does increase with the increased population density. _____

Section 44.4 Regulation of Population Size (continued)

EK 2.D.1, 2.D.3

SUMMARIZE IT

A volcano erupts and destroys an entire village nearby. Is the volcano eruption a density-independent or dependent factor? **Explain** your answer.

_____

_____

_____

EK 4.A.5, 4.B.3

Describe how competition for food could control population growth.

_____

_____

_____

Section 44.5 Life History Patterns, *pp. 846–849*

Essential Knowledge Covered:
2.D.1, 4.A.5

EK 2.D.1
Student Edition pp. 846–849

REVIEW IT

Describe the populations with *K*- or *r*-selection.

K-selection	*r*-selection

Which life history pattern do humans most closely follow?

_____

EK 2.D.1, 4.A.5
Student Edition pp. 846–847

USE IT

Identify if the statement pertains to an *r*- or *K*-strategist (r or K).

Often good at colonizing new habitats _____

Have a fairly long lifespan _____

Many offspring die before reproducing _____

Favor stable, predictable environments _____

EK 2.D.1

SUMMARIZE IT

Are all organisms either *K*- or *r*-strategists?

_____

_____

Section 44.6 Human Population Growth, *pp. 849–852*

Essential Knowledge Covered:
4.A.5, 4.A.6

EK 4.A.6
Student Edition p. 850

USE IT

Describe three problems a less-developed country might face if populations continue to climb in these regions.

_____

_____

_____

EK 4.A.5, 4.A.6

SUMMARIZE IT

People living in more-developed countries face a second type of overpopulation. Use the environmental impact equation to **describe** this second type of over-population.

_____

_____

_____

AP Reviewing the Essential Questions

These questions were posed in the *Biology* chapter opener (page 836). Answer them using the knowledge you've gained from this chapter.

1. How do environmental factors, including energy availability, affect the density and distribution pattern of a population?

_____

_____

_____

2. What can limit the size of a population? What are examples of density-independent and density-dependent factors?

_____

_____

_____

3. How can mathematic models be used to explain exponential and logistic growth of populations?

_____

_____

_____

45 Community and Ecosystem Ecology

As you work through your AP Focus Study Guide, keep this chapter's Big Ideas in mind:

FOLLOWING THE BIG IDEAS

 BIG IDEA 1 The pace of evolution is tied to the stability of an organism's environment.

BIG IDEA 2 Food chains and webs ensure that biomolecules cycle through the ecosystem.

 BIG IDEA 4 Communities differ in their species makeup and in their abiotic conditions.

Section 45.1 Ecology of Communities, *pp. 856–864*

Essential Knowledge Covered:
2.D.1, 2.E.3, 4.A.5, 4.A.6, 4.B.3

REVIEW IT

Match the vocabulary word to its definition.

EK 4.A.5, 4.A.6
Student Edition pp. 856–864

competitive exclusion principle	different species interacting with one another in the same environment
predation	the role an organism plays in its community
mimicry	the ability to resemble another species
community	the ability to blend into the background
resource partitioning	no two species can indefinitely occupy the same niche at the same time
habitat	when one organism feeds on another organism
character displacement	where a species lives and reproduces
camouflage	the tendency for characteristics to diverge when populations belong to the same community
ecological niche	apportioning of resources in order to decrease competition

EK 4.A.5
Student Edition p. 856

List two ways community structure can be compared.

_____

_____

Section 45.1 Ecology of Communities (continued)

USE IT

EK 4.A.5
Student Edition p. 857

Which is more diverse, a pond which has 30 wood frogs, 20 green frogs, and 10 bull frogs? Or a pond that has 80 wood frogs, 2 green frogs, and 5 bullfrogs. Why?

_____

_____

EK 2.D.1, 2.E.3, 4.B.3
Student Edition p. 856

Bats and bluebirds both live near fields and both eat the same types of moths and beetles. How are they able to occupy the same habitat?

_____

_____

_____

EK 2.E.3
Student Edition p. 862

Explain the difference between Batesian and Müllerian mimicry.

_____

_____

_____

SUMMARIZE IT

EK 2.D.1

Compare and contrast camouflage and mimicry.

_____

_____

_____

_____

Section 45.1 Ecology of Communities (continued)

EK 4.A.5

In a famous niche partitioning study, Joseph Cornell studied two species of barnacles living on the same rock. Using the illustration provided below, **describe** the niches and characteristics that allow two species to occupy the same rock.

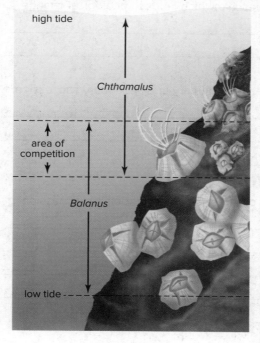

_____

_____

_____

_____

_____

_____

_____

Section 45.2 Community Development, *pp. 866–867*

Essential Knowledge Covered:
1.A.3, 4.B.4

Vocabulary

climax community
ecological succession
pioneer species

REVIEW IT

Identify the pioneer species and climax community in the drawing of ecological succession above.

_____

_____

Section 45.2 Community Development (continued)

EK 1.A.3, 4.B.4
Student Edition p. 866

USE IT

Describe an event which happens on a short timescale and an event which happens over a long timescale which leads to ecological change.

_____

_____

EK 4.B.4
Student Edition p. 866

Distinguish between primary and secondary succession.

_____

_____

SUMMARIZE IT

EK 4.B.4

Compare and contrast the three models of succession.

_____

_____

_____

_____

_____

_____

_____

Section 45.3   Dynamics of an Ecosystem, *pp. 868–878*

Essential Knowledge Covered:
2.A.1, 2.A.3, 2.D.1, 4.A.5, 4.A.6

REVIEW IT

EK 2.A.1, 4.A.5, 4.A.6
Student Edition pp. 868–870

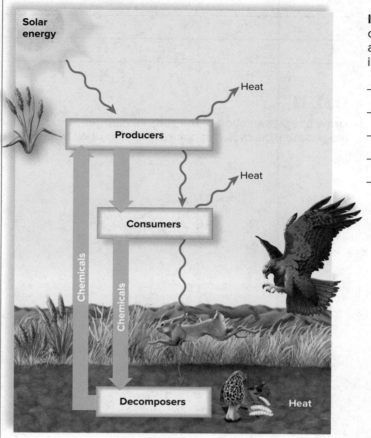

Identify the autotrophs, carnivores, herbivores, and detritivores in the illustration to the left.

_____

_____

_____

_____

_____

EK 2.A.1, 4.A.5, 4.A.6
Student Edition pp. 870–871

Using the organisms in the figure above, **draw** a grazing food chain and an ecological pyramid.

EK 4.A.5, 4.A.6
Student Edition pp. 873–874

List the four main biogeochemical cycles.

_____

EK 2.A.3, 4.A.6
Student Edition p. 874

Identify two greenhouse gases.

_____

Section 45.3 Dynamics of an Ecosystem (continued)

EK 2.A.3, 4.A.6
Student Edition p. 874

Explain the difference between the greenhouse gas effect, climate change, and global warming.

_____

_____

_____

_____

USE IT

EK 2.A.3, 4.A.6
Student Edition pp. 874–876

Draw a representation of the carbon, nitrogen, and phosphorus cycles. You may include biogeochemical cycles in one diagram, or draw a separate diagram for each.

Section 45.3 Dynamics of an Ecosystem (continued)

EK 4.A.5, 4.A.6
Student Edition p. 872, 878

Describe two ways that humans can interfere with the water cycle, and how this may affect other organisms.

_____

_____

_____

_____

SUMMARIZE IT

EK 2.A.1, 2.A.3, 4.A.5, 4.A.6

Imagine you are on a grassy plain in Africa. **Provide** some ideas of the types of producers, consumers, and decomposers which might be in that ecosystem. Where does the energy they need to reproduce and survive come from?

	Producers	Consumers	Decomposers
Examples			
Energy Source			

EK 2.D.1, 4.A.5, 4.A.6

Compare and contrast the phosphorus and nitrogen cycle.

_____

_____

_____

_____

_____

_____

_____

EK 4.A.5, 4.A.6

Describe what happens to a pond when runoff from fertilizer applied to a lawn introduces too much nitrogen into the system.

_____

_____

_____

_____

AP Reviewing the Essential Questions

These questions were posed in the *Biology* chapter opener (page 855). Answer them using the knowledge you've gained from this chapter.

1. How can human activities that alter biogeochemical cycles affect community composition and diversity?

   _____

   _____

   _____

2. What is the direction of energy flow among populations in an ecosystem? What might happen to the community if the producer base changes?

   _____

   _____

   _____

3. How do interactions among species, such as competition and predation, organize a community?

   _____

   _____

   _____

46 Major Ecosystems of the Biosphere

As you work through your AP Focus Review Guide, keep this chapter's Big Ideas in mind:

FOLLOWING THE BIG IDEAS

 BIG IDEA 4 Ecosystems and their keystone species are often adversely affected by formidable environmental changes.

Section 46.1   Climate and the Biosphere, *pp. 882–883*

Essential Knowledge Covered:
4.A.6

Vocabulary

climate
monsoon
rain shadow
topography

REVIEW IT

Fill in the vocabulary word or its definition.

Vocabulary Word	Definition
rain shadow	
	the prevailing weather conditions in a particular regions
	the physical features of the land
monsoon	

USE IT

EK 4.A.6
Student Edition p. 882

Describe two factors which dictate the climate of a region.

_____

_____

_____

SUMMARIZE IT

EK 4.A.6

It is close to the autumnal equinox and arctic winds have been blowing east. What might the climate be like in New York City? What might the climate might be like near the Equator?

_____

_____

_____

_____

_____

_____

EK 4.A.6

SUMMARIZE IT

Characterize the climate and vegetation of the following biomes.

Biome	Climate	Vegetation Type
alpine tundra		
arctic tundra		
chaparral		
deserts		
grasslands		
savannas		
shrublands		
taiga		
temperate deciduous forest		
temperate grasslands		
temperate rain forest		
tropical rain forest		

Section 46.3   Aquatic Ecosystems, *pp. 895–902*

Essential Knowledge Covered:
4.A.6, 4.B.4

REVIEW IT

EK 4.A.6
Student Edition pp. 895–902

Identify the following aquatic vocabulary words.

Areas of biological abundance found in warm, shallow tropical waters	
Where fresh water and salt water mix	
The concentration of pollutants as they move up the food chain	
The region of the shoreline that lies between the high and low tidal marks	
Offshore winds cause cold nutrient-rich waters to rise and displace warm nutrient-poor water	
A place where seawater is heated to about 350°C and percolates through the cracks at the bottom of the ocean	
The lack of upwelling which causes stagnation and climate patterns to change	

USE IT

EK 4.A.6
Student Edition p. 895

List three ways that wetlands benefit humans.

_____

_____

_____

EK 4.A.6
Student Edition p. 897

List the four zones of a lake, and provide an example of what lives in each zone.

_____

_____

_____

SUMMARIZE IT

EK 4.A.6, 4.B.4

Describe how ocean currents can change the climate and nutrients of an aquatic habitat.

_____

_____

_____

_____

_____

AP Reviewing the Essential Questions

These questions were posed in the *Biology* chapter opener (page 881). Answer them using the knowledge you've gained from this chapter.

1. What environmental factors determine the location and nature of terrestrial and aquatic biomes and ecosystems?

_____

_____

_____

2. How can changes in one ecosystem caused by natural events or human activity impact other ecosystems, especially if accompanied by the loss of keystone species?

_____

_____

_____

47 Conservation of Biodiversity

As you work through your AP Focus Review Guide, keep this chapter's Big Ideas in mind:

FOLLOWING THE BIG IDEAS

BIG IDEA 1   Humans impact evolution when they cause ecological stress.

BIG IDEA 4   The activities of human populations often unbalance ecosystems.

Section 47.1   Conservation Biology and Biodiversity, *pp. 906–907*

Essential Knowledge Covered:
1.A.2, 4.C.3, 4.C.4

Vocabulary

biodiversity
biodiversity hotspots
bioinformatics
community diversity
conservation biology
endangered species
genetic diversity
landscape
landscape diversity
threatened species

EK 1.A.2, 4.C.3, 4.C.4
Student Edition p. 906

REVIEW IT

Using the vocabulary on the left, **fill in the blanks** below.

_____ is the variety of all life on Earth. The study of maintaining biodiversity and natural resources is called _____. Species which are in peril of immediate extinction are called _____, and organisms that are likely to become endangered are called _____. Many scientists in this field employ _____ to analyze the vast amount of data biodiversity studies contain. There are three levels of biological organization in biodiversity: (1) _____, (2) _____, and (3) _____. _____ are regions of the world which contain a large concentration of species. Reducing fragmentation of the _____ is important in maintaining the biodiversity of these regions.

USE IT

Why is a small, isolated population more likely to become extinct than a larger, more connected population?

_____

_____

_____

Section 47.1   Conservation Biology and Biodiversity (continued)

SUMMARIZE IT

EK 1.A.2, 4.C.3, 4.C.4

Why do conservation biologists study genetic, community, and landscape diversity as well as just counting the numbers of a particular species?

_____

_____

_____

_____

EK 1.A.2, 4.C.3, 4.C.4

How can it be detrimental to only try to conserve a charismatic species, such as when opossum shrimp were introduced in Flathead Lake to try and boost the salmon population?

_____

_____

_____

_____

Section 47.2   Value of Biodiversity, *pp. 908–911*

Essential Knowledge Covered:
1.A.2, 1.C.3, 4.A.5, 4.C.3, 4.C.4

REVIEW IT

EK 4.A.5
Student Edition pp. 908–911

Explain the difference between the direct and indirect value of biodiversity.

_____

_____

_____

EK 4.A.5
Student Edition pp. 908–911

List two examples of each value of biodiversity in the table below.

Direct Value	Indirect Value

USE IT

EK 1.C.3, 4.C.4
Student Edition p. 909

Why is it important to maintain genetic diversity with a crop species?

_____

_____

_____

Section 47.2   Value of Biodiversity (continued)

EK 4.C.4
Student Edition pp. 909–911

Provide an example of how biodiversity is critical in each of the following areas:

biogeochemical cycles	
waste recycling	
soil stability	
climate	

SUMMARIZE IT

How might a high degree of biodiversity help an ecosystem function more efficiently?

EK 1.A.2, 4.C.3, 4.C.4

_____

_____

_____

_____

Section 47.3   Causes of Extinction, *pp. 911–916*

Essential Knowledge Covered:
1.C.1, 4.A.5, 4.A.6, 4.B.3, 4.B.4, 4.C.4

REVIEW IT

EK 4.A.5, 4.B.4
Student Edition pp. 911–915

Define the following threats to biodiversity.

habitat loss	
overexploitation	
pollution	
exotic species	
climate change	

Section 47.3   Causes of Extinction (continued)

USE IT

EK 4.A.5, 4.A.6
Student Edition p. 912

List two species humans have accidently transported from one environment to another which threaten biodiversity.

_____

_____

EK 1.C.1, 4.A.6, 4.B.3, 4.B.4
Student Edition pp. 911–913

How can introducing a new species to an environment it is not native to affect the other species living there?

_____

_____

_____

_____

EK 1.C.1, 4.B.4
Student Edition p. 913

Describe three types of pollutants and how they negatively affect biodiversity.

_____

_____

_____

SUMMARIZE IT

EK 1.C.1, 4.A.5, 4.A.6, 4.B.3,
4.B.4, 4.C.4
Student Edition p. 916

In the marine ecosystem off of the coast of California, sea otters are recovering from a genetic bottleneck caused by overharvesting by the fur trade. Sea otters eat sea urchins. Urchins are a main grazer on kelp, which provides a habitat for commercial and recreational fisheries. Seals and sea lions feed on fish. Orcas will eat otters but prefer sea lions.

Draw a diagram of the coastal food web. What will happen to the coastal fisheries the sea otter population is reduced by disease, as is a risk for populations in a genetic bottleneck?

Section 47.3 Causes of Extinction (continued)

EK 1.C.1, 4.A.5, 4.A.6, 4.B.3,
4.B.4, 4.C.4
Student Edition p. 911

A mass extinction is currently underway, undermining the biodiversity of our planet. Using your textbook, **find and record** the percentage species affected by each threat. What do all these reasons have in common?

Cause	Percentage
habitat loss	
exotic species	
pollution	
overexploitation	
disease	
Common cause:	

Section 47.4 Conservation Techniques, *pp. 916–918*

Essential Knowledge Covered:
4.B.4

REVIEW IT

Using the vocabulary on the left, **fill in the blanks**:

Vocabulary

edge effect
flagship species
keystone species
metapopulation
sink population
source population

A species that plays a fundamental role in the operation of a community is known as a

_____. These are different from _____ which are species that

evoke a strong emotional response in humans. In order to help conserve species, it is

important to understand that some species now live in _____ as a result of

habitat fragmentation. Within these populations, there is a _____, in which

one population produces an abundance of individuals, which may migrate to a

_____. Fragmentation leads to an increase in the _____, where the

characteristics the edges of a habitat are very different from the inside of that habitat.

USE IT

EK 4.B.4
Student Edition pp. 917–918

Refer to Figure 47.11 on page 918 of your textbook. **Describe** what happens to viable habitat when an environment becomes fragmented into smaller and smaller pieces.

_____

_____

_____

_____

Section 47.4 Conservation Techniques (continued)

EK 4.B.4
Student Edition pp. 917–918

SUMMARIZE IT

Describe the three key principals of restoration ecology.

_____

_____

_____

AP Reviewing the Essential Questions

These questions were posed in the *Biology* chapter opener (page 905). Answer them using the knowledge you've gained from this chapter.

1. How are human activities contributing to the endangerment and possible extinction of other species? What responsibility do we have for maintaining the Earth's biodiversity?

_____

_____

_____

2. What is the value of biodiversity to humans? Why does it matter if species become extinct?

_____

_____

_____